Understand Electronics

Understand
Electronics

Owen Bishop

Butterworth-Heinemann
Linacre House, Jordan Hill, Oxford OX2 8DP
A division of Reed Educational and Professional Publishing Ltd

℞ A member of the Reed Elsevier plc group

OXFORD   BOSTON   JOHANNESBURG
MELBOURNE   NEW DELHI   SINGAPORE

First published 1995
Reprinted 1996

British Library Cataloguing in Publication Data
Bishop, O. N.
  Understand Electronics
  I. Title
  621.381

ISBN 0 7506 2100 1

Library of Congress Cataloguing in Publication Data
Bishop, Owen.
  Understand electronics/Owen Bishop.
  p.   cm.
  Includes index.
  1. Electronics.   I. Title.
  TK7816.B543   95-15088
  621.381–dc20   CIP

Typeset by Keyword Typesetting Services Ltd, Wallington, Surrey
Printed in Great Britain

Contents

INTRODUCTION

This is a book for anyone who wants to get to know about electronics. It requires no previous knowledge of the subject, or of electrical theory, and the treatment is entirely non-mathematical. It begins with an outline of electricity and the laws that govern its behaviour in circuits. Then it describes the basic electronic components and how they are used in simple electronic circuits. Semiconductors are given a full treatment since they are at the heart of almost all modern electronic devices. In the next few chapters we examine a range of electronic sensors, seeing how they work and how they are used to put electronic circuits in contact with the world around them.

The methods used for constructing electronic circuits from individual components and the techniques of manufacturing complex integrated circuits on single silicon chips are covered in sufficient detail to allow the reader to understand the steps taken in the production of an item of electronic equipment. This is followed by an account of the test equipment used to check the finished product.

The next few chapters deal with the electronic circuits that are used in special fields and serves as an introduction to amplifiers, logic circuits, audio equipment, computing, telecommunications (including TV and video equipment) and microwave technology. Then we look at two other fields in which electronics plays an ever-increasing role – medicine and industry. Throughout, the descriptions are intentionally aimed at the non-technical reader.

Finally, we outline some of the current research in electronics and point the way to future developments in this technology.

1

Electrons and electricity

Electricity consists of electric charge. Though electricity has been the subject of scientific investigations for thousands of years, the nature of electric charge is not fully understood, even at the present day. But we do know enough about it to be able to use it in many ways. Using electric charge is what this book is about.

Electric charge is a property of matter and, since matter consists of atoms, we need to look closely at atoms to find out more about electricity. But, even without studying atoms as such, we are easily able to discover some of the properties of electricity for ourselves.

The simplest way to demonstrate electric charge is to take a plastic ruler and rub it with a dry cloth. If you hold the ruler over a table on which there are some small scraps of thin paper or scraps of plastic film, the pieces jump up and down repeatedly. If you rub an inflated rubber balloon against the sleeve of your clothing then place it against the wall or ceiling, for a while, the balloon remains attracted to the wall or ceiling, defying the force of gravity. The electric charge on the plastic ruler or wall is creating a force, an **electric force**. In effect, the energy of your rubbing appears in another form which moves the pieces of paper, or prevents the balloon from falling.

You can also charge a strip of polythene sheet (cut from a plastic food-bag) by rubbing it. If you charge two strips, then hold them apart (Figure 1.1) and then try to bring them together, they are repelled by each other. As you try to push them together, their lower ends diverge, spreading as far away from each other as they can. On the other hand, if you charge a strip of acetate sheet (cut from a shirt-box) by rubbing it and bring it toward a charged polythene strip, the two strips attract each other. If they are allowed to, their lower ends come together and touch. From this behaviour, we reason that the charge on an acetate strip must be of a different kind from that on a polythene strip. Simple demonstrations such as these show that:

- There are two kinds of electric charge
- Like kinds of charge repel each other
- Opposite kinds of charge attract each other.

1

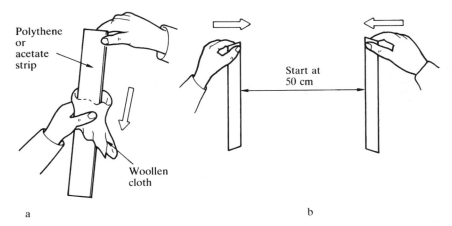

Polythene or acetate strip

Woollen cloth

Start at 50 cm

a   b

Figure 1.1   *(a) Charging a plastic strip (b) Bringing the charged strips together*

The discovery of electricity

Electricity takes its name from the Greek word *elektron*, the name of the resinous solid known as amber. The ancient Greeks had discovered that, when a piece of amber is rubbed with a soft cloth, it becomes able to attract small, light objects to it. We say that it has an **electric charge**.

More electric attraction

You may have noticed this effect in the shower. The fine spray of water droplets charges your body and the shower curtain, but the charges are opposite. There is an attractive force between your body and the curtain. The curtain billows inward and clings to your body.

Electric charge and atoms

Now we are ready to link the basic facts about electric charge to what is known about the structure of matter.

Research has shown that atoms are built up of several different kinds of atomic particle. Most of these occur only rarely in atoms but two kinds are very common. These are electrons and protons. Although protons are about 2000 times more massive than electrons, protons and electrons have equal electric charges. The charge of an electron is opposite in its nature to the charge on a proton; these are the two kinds of electric charge mentioned above. The charge on an electron is said to be negative and that on a proton to be positive, but this is simply a convention. There is nothing positive on a proton which is 'missing' or 'absent' from an electron. The two terms merely imply that positive and negative charges are opposite.

We have said that electrons and protons carry equal but opposite charges. If an electron combines with a proton, their charges cancel out exactly and an uncharged particle is formed – a neutron. Since neutrons have no charge, they are of little interest in electronics.

Atomic structure

All atoms are composed of electrons and protons (ignoring the other rare kinds of particle). The simplest possible atom, the atom of hydrogen, consists of one electron and one proton. The proton is at the centre of the atom and the electron is circling around it in orbit (Figure 1.2). With one unit of negative charge and one of positive charge, the atom as a whole is uncharged. Since the electron is moving at high speed around the proton, there must be a force to keep it in orbit, to prevent it from flying off into space. The force that holds the electron in the atom is the attractive electrical force between oppositely charged particles, which we demonstrated earlier. It acts in a similar way to the attractive force of gravity, which keeps the Moon circling round the Earth, and the planets of the solar system circling round the Sun.

Other atoms

There are more than a hundred different elements in nature, including hydrogen, helium, copper, iron, mercury and oxygen, to name only a few. Each

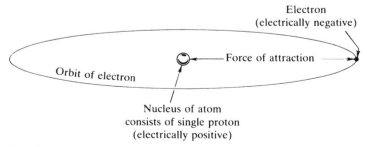

Figure 1.2 *The structure of a hydrogen atom*

element has its own distinctive atomic structure, but all are based on the same plan as the hydrogen atom. That is to say, there is a central part, the nucleus, where most of the mass is concentrated, which is surrounded by a cloud of circling electrons. However, atoms other than hydrogen have more than one proton and also some neutrons in the nucleus at the centre of the atom. The positive charge on the nucleus is due to the protons it contains. The electron cloud contains a number of electrons to equal the number of protons in the nucleus. In this way the positive charge on the nucleus is exactly balanced by the negative charges on the electrons and the atom as a whole has no electric charge.

The electrons are in orbits at different distances from the nucleus. These orbits are at definite fixed distances from the nucleus and there is room for only a fixed number of electrons in each orbit.

Atomic dimensions

The orbit of the electron of a hydrogen atom is about one ten-millionth of a millimetre in diameter. If the atom was scaled up so that its nucleus (the proton) was 1 mm in diameter, the orbiting electron would be a tiny speck about 120 m away. The interesting point is that the electron and proton take up very little room in the atom. So-called 'solid' matter is mostly empty space. The tangible nature of matter is not due to it consisting mostly of firm particles. Instead, it is due to the strong electrical forces between atomic particles and the forces between adjacent atoms, which hold the atoms more-or-less firmly together. There is more about the structure of matter on p. 7.

Electric fields

When an object is charged there is an electric field around it. This is a force field which makes charged objects move when they are in the field. Another more familiar force field is gravity, which affects us everywhere and at all times; but gravity is only attractive, it does not repel.

Figure 1.3a shows how we imagine the field around an electron. The lines of force show the way a positive charge moves when placed in the field; it moves towards the electron. Although lines of force are strictly imaginary (just as the lines of latitude and longitude on the Earth are imaginary) it helps to think of them as if they are like rubber bands under tension. This gives them two properties:

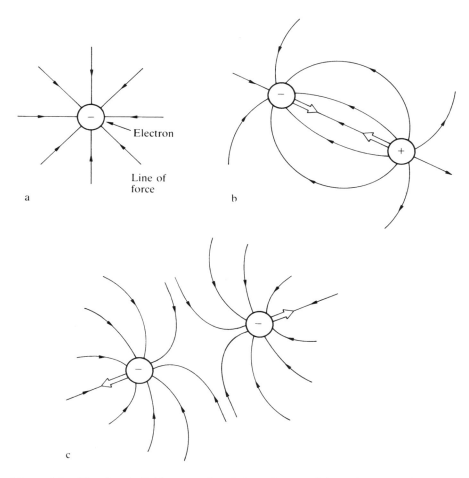

Figure 1.3 *The electric field surrounding (a) an electron (b) an electron and a proton (c) two electrons. The broad arrow indicates the motion of the particles*

- They tend to be as short as possible
- They tend to be as straight as possible.

In Figure 1.3b we see the fields of the electron and proton combining to give lines running from the proton to the electron. Making the lines as short as possible creates forces acting on the electron and proton, drawing them together until they meet. They are attracted to each other. In Figure 1.3c we have two electrons each with its own field. Another property of lines of force is that they can not cross. So the fields become distorted, as shown. But lines of force tend to become straight and, for this to happen, the electrons are forced to move further apart.

Charge and energy

We can now begin to understand what happens when we charge a strip of plastic by rubbing it with a cloth. Although there is normally a fixed number electrons circling around the nucleus of an atom, the electrons in the outer orbits are less strongly attracted to the nucleus than those closer to it. Rubbing the plastic with a cloth provides energy (derived from our muscles) to overcome the attractive forces between the nucleus and the outer electrons. Depending on the nature of the plastic, we may remove electrons from some of the atoms in the molecules of the plastic and attach them to the atoms in the molecules of the cloth. Removing electrons leaves the plastic with excess positive charge; collecting electrons on the cloth gives it negative charge. When we pull the cloth away from the plastic at the end of the rubbing process, the strip and cloth are oppositely charged, so they are attracted to each other. We need to use more muscular energy to pull them apart now than if they were uncharged.

When the strip and cloth have been separated, they remain charged until charged molecules in the air are attracted to them. The plastic which, lacking electrons, is positively charged, attracts any negatively charged molecules that happen to be in the surrounding air. The electrons on these are passed across to the charged atoms of the plastic, so gradually discharging them. A similar process discharges the cloth.

Charging other substances

Many kinds of substance are charged when rubbed with a cloth: possible substances include different kinds of plastic, rubber, and glass. Exactly what happens depends on which substance is rubbed with which kind of cloth. On this page we describe how the substance becomes negatively charged and the cloth becomes positively charged. But with a different substance or a different kind of cloth, charging may occur in the opposite direction.

Conduction

Substances such as plastic, wool, glass, and rubber can be charged because there is no quick way that electrons removed from an atom can be replaced or that excess electrons can be got rid of. The charged atoms are isolated from each other and from the surroundings, except when, for example, a charged molecule is attracted from the air, or the substance is brought into contact with an oppositely charged surface. Substances of this kind, in which charges stay in a fixed place, are known as **non-conductors** or **insulators**. As well as those mentioned above, this class of substance includes wood, paper, ceramics, pure water, asbestos and air.

The other major class of substance comprises the **conductors**. These are mostly elements such a silver, gold, copper, tin, lead, and carbon. As can be seen from this list, the majority of them are metals. Conductors also include alloys of metals (for example, brass and solder) and solutions of salts, but first we describe conduction as it occurs in metals. Figure 1.4 illustrates the structure of a typical metallic conductor: a block of copper. Metals have the structure of crystals, the individual atoms are arranged in a regular three-dimensional array known as a matrix or lattice. They are held in position by forces (not electric) existing between each atom and its neighbours. Each atom has electrons circling its nucleus but, in metals, the outer atoms are only weakly attracted to the nucleus. They are able to leave the atom and wander off at random into the spaces within the lattice, not being attached to any particular atom. When these electrons move, negative charge moves from one part of the copper block to another. We say that the electrons are **charge carriers**.

Figure 1.5 shows one way in which the electrons may be made to move in a more orderly fashion. The copper block is connected to an electric cell (p. 11) by copper wires. There is an electric field between the positive and negative terminals of the cell. The way this field is generated is described later but, for the moment, consider there to be lines of force running from the positive terminal to the negative terminal. Before the cell is connected to the block, the field lines run directly between the terminals, through the air. When the cell is connected by wires to the copper block, the field lines mostly become bunched together, and run through the wires and block, instead of running

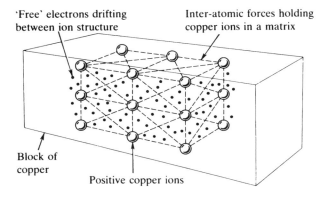

Figure 1.4 *The arrangement of copper atoms in the lattice of metallic copper*

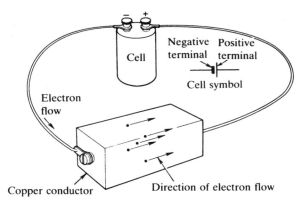

Figure 1.5   *The flow of electrons in a conductor*

through the air. Thus there is an electric field running from the positive terminal, through the wire on the right, through the block, through the wire on the left and back to the negative terminal. Any charged particles in that field will move, if they are free to do so. If the block was made of plastic or some other non-conductor, charged particles would not be free to move. In a conductor such as copper, the electrons wandering between the atoms are very free to move. Now, instead of wandering randomly, they all flow in one direction through the block (and wires), repelled by the negative terminal of the cell and attracted toward its positive terminal. We have an **electric current**. This is just what an electric current is – a mass flow of electric charge from one place to another.

As electrons move through the copper block and wires, those reaching the positive terminal of the cell pass into the cell. They are replaced by electrons coming from the negative terminal. Most of electronics deals with the flow of electric currents. We are not so interested in the stationary (or static) charges built up by rubbing non-conductors, but there is one instance in which static charges are really important and we must take special precautions to eliminate them, as explained on p. 78.

Electromotive force

When a charged particle is in an electric field it is subject to a force, making it move if it is free to do so. This force is known as **electromotive force**, often referred to briefly as e.m.f. In Figure 1.5 there is an e.m.f. between the terminals of the cell, making current (charge carriers) flow through the wires and copper block.

Current through a solution

The charge carriers in a solution in water are ions of the dissolved substance. For example, when copper sulphate ($CuSO_4$) is dissolved in water, its molecules break up into two ions: copper (Cu) and sulphate (SO_4). When they break up, or ionize, the sulphate ion takes two electrons from the copper atom, leaving it positively charged (Cu^{++}). This makes the sulphate ion negatively charged (SO_4^{--}). Figure 1.6 illustrates the flow of charge carriers between two copper rods immersed in the solution and connected externally to a cell (not shown). The copper ions are attracted toward the negative rod and are there discharged by electrons which have come from the negative terminal of the cell. The discharged copper ions are deposited on the rod as a bright reddish layer of copper. Copper atoms in the positive rod dissolve in the water, each losing two electrons which flow to the positive terminal of the cell. This copper rod gradually becomes thinner. The sulphate ions are attracted toward the positive supply but do not become discharged, so they are not charge carriers.

Although salts form ions when dissolved in water, making the salt solution a conductor, pure water does not form ions, and so it is a non-conductor.

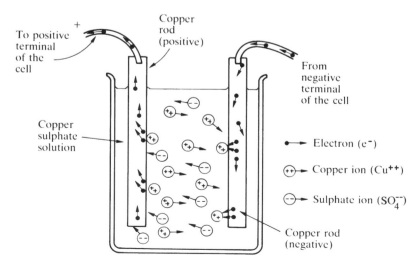

Figure 1.6 *The flow of electric charge through a solution of copper sulphate; the cell which provides the current is not shown*

Current through a gas

Figure 1.7 shows charge flow through neon gas at low pressure. There is two-way conduction. Electrons flow from the negative plate (negative electrode) to the positive plate (positive electrode). On their way, they strike neon atoms and knock electrons out of them. This creates more electrons to act as negative charge carriers. The neon atoms which have lost electrons become positive ions, and act as positive charge carriers. The energy from the moving carriers excites many of the neon atoms. Excited atoms later lose this energy, which then appears in another form, that of reddish light.

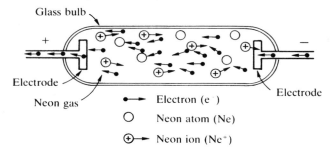

Figure 1.7   *The flow of electric charge through a neon lamp*

2
E.m.f. and potential

The terms **electromotive force** and **potential** are often used as if they mean the same thing. Although they have some features in common, they are not quite identical. This chapter helps you to understand the difference between them.

A source of electromotive force

The source of electromotive force (e.m.f.) in Figure 1.5 is an electric cell. In Chapter 1 we explained how the e.m.f. of the cell gives rise to a field between the terminals of the cell, and how the force produced by this field drives electrons around the circuit from the negative terminal to the positive terminal of the cell.

There are many types of cell but, to illustrate the principles of the way cells work, we will look as a simple wet cell (Figure 2.1). The cell has two **electrodes** made from conductors such as zinc, copper or carbon. In the example, one

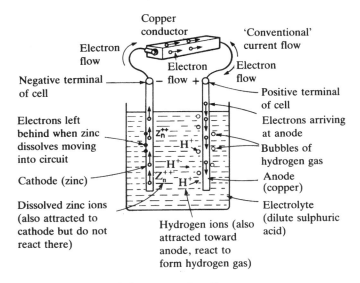

Figure 2.1 *Producing an e.m.f. from a simple cell*

electrode is made of zinc and the other of copper. The electrodes are immersed in a solution, known as an electrolyte. In the example the electrolyte is dilute sulphuric acid. This is ionized (like the copper sulphate solution described on p. 9) into hydrogen (H^+) and sulphate (SO_4^{--}) ions. Some of the zinc atoms dissolve in the acid, forming zinc ions (Zn^{++}). The electrons freed when the zinc dissolves are left on the zinc electrode, giving it a negative charge. The action is driven by chemical energy released as a result of the zinc going into solution. It continues for a while until the increasing negative charge on the zinc electrode attracts zinc ions back to the electrode in such quantities that no further dissolving occurs. We are left with the zinc electrode being negatively charged with respect to the electrolyte and to the copper electrode. In an electric cell, the negative electrode is called the **cathode**, while the positive electrode is called the **anode**. There is an e.m.f., creating an electric field between the cathode and anode.

If the terminals of the cell are connected by wires to a block of copper or other metallic conductor, the electric field between the electrodes forces electrons to pass from the zinc cathode to the copper anode. The cell begins to supply an electric current. Now that electrons are flowing out of the cathode into the wires, the cathode becomes less negatively charged than before and attracts zinc ions less strongly. This means that more zinc is free to dissolve, producing a further supply of electrons. Electrons that have flowed through the wires reach the anode. There they attract H^+ ions from the electrolyte. They discharge the hydrogen ions, which form bubbles of hydrogen gas.

As long as there is a wire or other electrical connection between the terminals of the cell, the action of the cell continues as described above. The e.m.f. makes a current of electrons flow from the cathode to the anode. The cathode is steadily eaten away as the zinc dissolves (this is what is driving the action, converting chemical energy into electrical energy) and bubbles of hydrogen gas rise steadily from the anode. The process continues until the zinc is completely dissolved.

Other sources of e.m.f. include electrical generators, deriving their energy from coal, oil, atomic power, water turbines, the wind, or the waves. Solar panels, such as are used on satellites, generate e.m.f. from the energy of sunlight (p. 124).

Which way does the current flow?

Figure 2.1 shows a flow of electrons (a **current**) from the negative terminal of the cell to the positive terminal. Yet we are accustomed to thinking of current as flowing from positive to negative. Most of the diagrams in this book show it flowing in that direction. Current flowing from positive to negative is known as **conventional current**. It is just a convention that this is the direction in which it flows.

But, when we look more closely at the charge carriers themselves, we often find that, as in Figure 2.1, the actual flow is that of negative charge carriers (electrons) from negative to positive.

Usually it is more convenient (because everyone thinks that way) to regard current as flowing from positive to negative. But sometimes, for example when we are looking at the way transistors work (p. 83), it is more helpful to follow which way the actual charge carriers are moving.

Other cells

The wet cell described above is seldom used because it possesses several disadvantages. The acid electrolyte is a corrosive substance and makes the cell hazardous to handle. The fact that it generates hydrogen gas means that the cell can not be sealed to contain the acid safely. Hydrogen gas is explosive, so this is another source of danger. Other types of cell are available which work on the same principles, converting chemical energy to electrical energy, but with greater safety and with the ability to produce larger current.

Most other types of cell are dry cells, in which the liquid electrolyte is replaced by a stiff paste. In all types of cell the electrodes are of different materials and one of the electrodes may be shaped to become the container of the cell. For example, in the commonly-used zinc-carbon cell (the sort we use in an electric torch), the cylindrical container is made of zinc and is the cathode. The anode is a carbon rod running down the axis of the cell. A variety of dry cells is available having different features:

Zinc-carbon: the typical 'torch cell', the cheapest type of cell; liable to leak when it is old.

Zinc chloride: can deliver large currents, produces more power; low leakage.

Alkaline: high currents, holds about three times as much charge as a zinc-carbon cell.

Silver oxide: low power but long-lasting; made as button-cells for watches and calculators.

Mercury oxide: similar to silver oxide.

Lithium: low current for long times or high current for short times, store large amounts of charge (about three times as much as alkaline cells) and remain in working condition for many years; often used as a back-up supply for computer memory. Recent research has been directed to manufacturing flexible lithium cells, cased in plastic, that can be bent to fit into confined or oddly-shaped spaces in portable equipment.

Although manufacturers usually state that ordinary dry cells are unsuitable for recharging, it has been shown that their life may be considerably extended if the recharging is done in a certain way. Both zinc chloride and alkaline cells may be recharged in this way and the chargers may also be used with nickel-cadmium cells (below). The process consists of slow charging periods interspersed with periods during which the cells are partially discharged. Gradually the charge builds up, restoring the full voltage to the cells in 8 to 12 hours. A cell may be recharged as many as 20 times. World-wide, a total of nearly 12 billion dry cells are discarded each year; the introduction of the new battery chargers will help to reduce the considerable amount of waste and environmental pollution that this implies.

Types of cell intended for re-charging include:

Nickel-cadmium (NiCad): store less charge than the zinc-carbon cells, but can deliver a high current. One disadvantage is that, if they are not allowed to discharge fully before recharging, they 'remember' the level at which they were recharged. Subsequently, they cease to deliver current when they have been discharged to the 'remembered' level, reducing their capacity appreciably.

Lead-acid: very high current, high charge.

Nickel metal-hydride (NiMH): cells have several advantages over the NiCad type, one of which is that they are able to store 30–50% more energy in a given volume. They do not exhibit the 'memory effect' so it is not necessary to discharge them completely before re-charging. They are particularly suited to use in electrically-powered vehicles.

Cells and batteries

A cell is a single unit. A battery consists of several cells connected together. Usually the cells of a battery are joined cathode-to-anode as in Figure 2.2. The advantage of joining cells in this way is that the e.m.f. of the battery equals the total of the e.m.f.s of the individual cells. The electric field is that much stronger and so is the current produced.

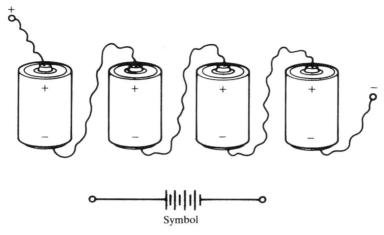

Figure 2.2 *Four cells joined to make a battery*

New batteries

With the increasing use of portable electronic equipment, such as cellphones and lap-top computers, there is an incentive to develop power sources suited to their special needs. The usual aim is to combine high capacity with small physical size. Other desirable features are a constant voltage up to the time of complete discharge and the possibility of recharging. Polymer batteries have their anode made from foil of a conductive polymer called polypyrrol. The battery has the desirable properties of lithium batteries (p. 13). For lap-tops, work is being done on zinc–air batteries which are able to store a large amount of power in small size. Such batteries will be able to power a lap-top computer for 10 to 20 hours before needing recharging.

Perhaps the most interesting development is the smart battery. This has a small integrated circuit (p. 136) in it which links up with a microprocessor in the smart charger. It should be explained that for most effective battery operation and maximum battery life, charging needs to be performed according to a carefully controlled routine. The charging current must be adjusted continually according to the amount of charge already in the battery and the temperature of the battery. Sensors in the battery report on battery status to the microprocessor in the charger and the optimum charging program is selected automatically.

Measuring current, charge and e.m.f.

The fundamental electrical unit is the unit of current, the **ampere**. This is often known as an 'amp' for short, and its symbol is A. The ampere is measured by measuring the force between two coils each carrying the same current. We do not need to be concerned with the size of the force or the dimensions of the coils. It is enough to know that current can be measured in a purely physical way. Having defined the ampere, we define the unit of charge, the **coulomb** (symbol, C) by saying that if a current of 1 A flows for 1 second, the amount of charge carried is 1 C. A coulomb is a relatively large amount of charge when compared with the charge on an electron. The charge on an electron is only 1.6×10^{-19} C (one sixteen trillionths of a coulomb). This indicates that there must be an unimaginably large number of electrons flowing through the element of an electric kettle when the current through it is, say, 2.5 A.

Another important unit is the unit of **electrical potential**. This is the unit used for expressing e.m.f. To understand this we will look first at a similar idea, the idea of gravitational potential. If we lift a brick up from the ground, we are acting against gravity. We are doing work, or expending energy to lift the brick. If we let the brick go, it drops down again; our energy reappears as the motion of the brick, the noise made as it hits the ground, and possibly the energy expended as it shatters. While we are holding the brick still, above ground, it does not appear to have any extra energy. But it actually has potential energy, the energy stored in it because of its position above ground. The amount of potential energy (p.e.) stored in the brick depends on two things: the mass of the brick and the height above ground to which we lift it. The higher we lift it, the greater its potential energy, the more noise it will make when it hits the ground, the more likely it is to dent the ground, and the more likely it is to shatter on impact. If we take several identical bricks and lift them to different heights, the different heights are associated with different amounts of potential energy (per brick). Each point above the ground has a different potential, beginning with zero potential at ground level, and increasing with the distance above ground. We could say that every point has a potential depending on how much energy we need to lift a brick to that point.

The same kind of thinking applies to an electric field. Consider a positively charged object A (Figure 2.3), fixed in space with an electric field around it (not shown). Take another positively charged object B, which is a long way from A, at the position marked B_1, so that the effect of electrical repulsion is very small. Now move B toward A. We must use energy to do this because both are positively charged. A repels B and we use energy to overcome this force. If we let B go free, it travels back until it is once more a long way from A. The energy stored in B when we force it toward A is recovered as the motion of B away from A.

If we bring B to a point X (B_4 in Figure 2.3) and hold it there, it does not appear to have any energy, but it has potential energy. We can say that:

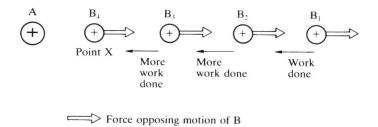

Figure 2.3 *Moving charged particle B up close to A requires work to overcome the opposing force. B_1, B_2, B_3 and B_4 are successive positions of B*

The electrical potential at X is measured by the amount of energy needed to move B to point X from a great distance.

It makes a better definition if we say that:

The electrical potential at any point in space is the amount of energy (measured in joules) needed to move a unit charge (1 C) to that point from a great distance.

So the unit of electrical potential is joules per coulomb (symbol J/C), in other words, 'How many joules are needed to move a coulomb of charge to that point?' We use the idea of potential so often that we give the unit 'joules per coulomb' a special name of its own, the **volt**, symbol V.

Potential difference

It is not really a practical task to measure the potential at a point, for example, the potential at the top right-hand corner of this page in the book. We would have to start at some point (if it exists) in outer space, far from any stars or galaxies, and measure how much work we did in bringing a unit charge (if we had the means of handling it) from there to the book. Much more important is to be able to measure the difference between the potentials of two nearby points, such as two points in the same electronic circuit. We refer to this as **potential difference**. The idea of potential difference (shortened to p.d.) can be explained by comparing it with potential due to gravity. When we climb a hill, we may like to know how far up it is from bottom to top. We are only interested in the difference in heights, not the actual heights of bottom and top as measured from sea level or from the centre of the Earth.

Consider the electric field between the anode and cathode of a cell. If we place a unit positive charge (1 C) near the cathode and move it toward the anode, we have to use energy to overcome the attraction toward the cathode and the repulsion from the anode. Moving the charge toward the anode gives it more potential energy. Its potential is greater there than when it is beside the cathode. There is a potential difference between the two points. For a typical dry cell, the potential difference (or p.d.) is about 1.5 V. That is to say, we need to expend 1.5 J of energy to move a 1 C positive charge from the cathode to the anode. Conversely, if a 1 C positive charge is placed at the anode and is allowed to flow (through a wire, for example) from anode to cathode, 1.5 J of energy is released. What becomes of this energy we will see later.

Units, multiples and sub-multiples

The electrical units have a system of multiples and sub-multiples. In electronics we must often use sub-multiples for current and potential:

1 A	= 1000 mA (milliamps)	1 V	= 1000 mV (millivolts)
1mA	= 1000 μA (microamps)	1mV	= 1000 μV (microvolts)

For the distribution of the electric mains we also use multiples of potential:

1000 V = 1 kV (kilovolts)

Potential and voltage

The electrical potential at a point or the p.d. between two points is measured in volts. 'Potential' and 'potential difference' or 'p.d.' are the strictly correct terms, and we use them as often as we can in this book. The terms 'voltage' and 'voltage difference', which mean the same thing, are used in a rather more practical sense, particularly in the context of electrical (as opposed to electronic) descriptions. There are instances where these terms seem to sound better than 'potential' and 'p.d.' and then we use them.

E.m.f. and potential

To sum up the main points of this chapter, e.m.f. is a force due to an electric field produced by some kind of generator such as a cell. The e.m.f. of a cell is the result of the chemical activities taking place inside it. Without a practical generator of some kind, there can be no e.m.f. Potential is a property of each and every point in space, depending how much energy is required to bring unit charge to that point. It is a much more theoretical concept, though potential difference may sometimes be the result of an e.m.f. Both are measured in volts.

For more insight into e.m.f. and potential, see p. 20.

3

Resistance

When electrons or other charge carriers flow through a conductor, they move through the lattice and pass close to the atoms of the conductor. They may actually collide with them or at least lose energy to them when they come within range of the electrical fields around the atoms or their nuclei. Whatever happens, the electrons lose energy. We say that the conductor offers **resistance** to the flow of current. At the same time, the atoms of the conductor gain energy. The effect of gaining energy is usually to make the atoms vibrate slightly about their usual fixed positions in the lattice. The motion of one atom is transferred to other atoms in the lattice, so that all of the atoms are in slight motion. This type of motion is a form of thermal energy, usually known as heat. When a current passes through a conductor, the conductor becomes warm or even very hot. In the case of the filament of an electric lamp, it becomes so hot that it glows with visible light. Some of the electrical energy has been converted to heat and some to light energy.

When a charge carrier begins its journey through a conductor it has potential energy, depending on the electrical potential at the point it starts from. During its passage though the conductor it loses some of this potential energy, which is used to heat the conductor. The loss of energy gives rise to a potential difference (p. 17) between one end of the conductor and the other. The amount of energy lost is proportional to the number of charge carriers passing through the conductor, in other words, to the current. We can say that:

The potential difference between two ends of a conductor of a given material is proportional to the current.

This statement is a version of **Ohm's Law**, first stated by Georg Ohm. This law can be put into mathematical form. If a current I is passing through a conductor and there is a p.d. V between its ends, and we calculate R, where:

$$R = \frac{V}{I}$$

we find that, for any given conductor, R is constant. R is called the **resistance** of the conductor. If I is expressed in amps and V in volts, the resistance is in ohms.

For example, a conductor is connected to a battery which has a p.d. of 12 V between its terminals. It is found that a current of 4 A is passing through the conductor. What is its resistance?

In this example, $V = 12$ and $I = 4$ so:

$$R = V/I = 12/4 = 3$$

The resistance is 3 ohms. Instruments for measuring p.d.s and currents are described in Chapter 16.

The **ohm** is the unit of resistance and its symbol is Ω. It has as multiples, the kilohm ($1 \text{ k}\Omega = 1000 \text{ }\Omega$) and the megohm ($1 \text{ M}\Omega = 1000 \text{ k}\Omega$).

The resistance of any given conductor also depends on the nature of the conductor. It depends on the size and shape of the conductor and the material from which it is made. Some conducting materials have a bigger supply of charge carriers than others, which makes it easier for them to conduct a current. Examples of good conductors are copper and aluminium.

V and *V*

We use the symbol V, in upright (Roman) capitals for the unit of p.d., the volt.
We use the symbol *V*, in slanting (italic) capitals for the value of any specified p.d.

Resistors

All conductors except superconductors (see p. 27) offer resistance to electric current. In electronics we use special components, called **resistors**, to provide a given amount of resistance as required in the circuit. Figure 3.1 shows a typical resistor, as well as the symbols used for it in circuit diagrams. Various types of resistor are available, one widely used type being the metal film resistor. This consists of a rod of insulating ceramic coated with a metal film. This film forms a spiral track running from one end of the resistor to the other. It makes contact with the terminal wires at each end. The resistance of the resistor

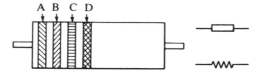

Figure 3.1   *A resistor showing four colour-coded bands and resistor symbols*

depends upon the length and the width of the track, and upon the metal it is made from.

Resistors can be made to a very high degree of precision, but precision resistors are unnecessary for most electronic circuits. Low-cost mass-produced resistors are suitable for most purposes. Resistors are manufactured with a quoted degree of precision, known as **tolerance**. If a resistor has a tolerance of 10%, for example, we know that the actual resistance is within 10% of its nominal value (the value marked on the resistor). For example, if a resistor is marked with a nominal value, such as 33 Ω and its tolerance is indicated as 10%, it means that the actual value of the resistor may be any value within 10% of 33 Ω. Since 10% of 33 Ω is 3.3 Ω, the value is between:

$$33 - 3.3 = 29.7\,\Omega$$

$$\text{and } 33 + 3.3 = 36.3\,\Omega$$

If we regard a nominally 33 Ω resistor with 10% tolerance as being close enough for circuit-building, there is no point in stocking resistors with other nominal values in the range 29.7 Ω to 36.3 Ω. The next useful lower value is 27 Ω, for resistors with this nominal value may lie between 24.3 Ω and 29.7 Ω. The range of 27 Ω resistors just touches the range of 33 Ω resistors. Similarly the next useful value above 33 Ω is 39 Ω, with a range from 35.1 Ω to 42.9 Ω.

To minimize wasteful overlapping of actual resistor values, resistors are made in a series of preferred values, known as the E12 series. In this series there are twelve basic values:

$$10, 12, 15, 18, 22, 27, 33, 39, 47, 56, 68, 82.$$

The series continues with twelve values ten times the above: 100, 120, 150, . . . And then it continues with twelve more values: 1000, 1200, 1500, . . . (or 1k, 1.2k, 1.5k, . . .)

This scheme is repeated decade by decade up to the maximum practicable value, which is usually 10 MΩ.

There is also a set of smaller values: 1.0, 1.2, 1.5, . . .

In this way the E12 series can provide resistors with 10% tolerance to completely cover the range 1 Ω to 10 MΩ.

Certain circuits may require resistance to be more closely fixed, in which case we can use resistors with 5% tolerance. With 5% tolerance there are small gaps in the ranges covered by adjacent values in the E12 series. If we need a resistor of 24 Ω, for example, this is above the range of 5% 22 Ω resistors yet below the range of 5% 27 Ω resistors. Quite often we just use a 22 Ω or 27 Ω resistor and the circuit works just as well but, if resistance is really critical, we can use resistors from the E24 series. This has all twelve values from the E12 series and an additional twelve in-between values. The E24 series has these additional values:

$$11, 13, 16, 20, 24, 30, 36, 43, 62, 75, 91.$$

The series repeats in each decade as with the E12 series. In the example, we would choose a 24 Ω resistor from the E24 series. For even greater precision, though at greater expense, resistors are made with 2% and 1% tolerance in the E48 and E96 series.

Colour codes

The resistance of a resistor is indicated by a number of coloured bands, the **colour code**, painted on the resistor, as shown in Figure 3.1. Two main systems are used, the four-band system and the five-band system. The four-band system uses the first three bands to indicate resistance. The colours represent numbers according to the following code:

Colour	Number
Black	0
Brown	1
Red	2
Orange	3
Yellow	4
Green	5
Blue	6
Violet	7
Grey	8
White	9

The first two bands (A and B in Figure 3.1) indicate the first two digits of the value. The third band (C) indicates a multiplier, specified as a power of 10. For example, if the resistance is 6800 Ω, the first two bands are blue and grey. This indicates 68. To obtain the actual resistance, this value must be multiplied by 100, or 10^2. The power of 10 required is 2 and red is the colour code for 2, so the three bands are:

blue, grey, red

Another example: if a resistor is marked with yellow, violet and yellow bands (in that order), the first two bands indicate 47. The multiplier is 10^4, which is 10 000. The value is $47 \times 10\,000 = 470\,000\,\Omega$, which is equivalent to 470 kΩ. This system may seem complicated at first but, in the E12 system, there are only ten possible pairs of colours for the two digits (brown/black, brown/red, brown/green, . . .) and these are soon learnt.

For resistors of less than 10 Ω, we use two sub-multipliers:

silver means × 0.1 and gold means × 0.01

For example, if the bands are green, blue, gold, the resistance is $56 \times 0.01 = 0.56\,\Omega$.

Resistors as p.d. producers

When a current passes through a resistor the charge carriers lose potential (p. 20). There is a potential difference between one end of the resistor and the other. The size of the potential difference is given by the Ohm's Law equation on p. 20. For this reason we can think of a resistor as a 'current to voltage' converter.

The colour codes for tolerance are:

Colour	Tolerance ($\pm\%$)
Silver	10
Gold	5
Red	2
Brown	1

The five-band colour code in required for marking high precision resistors in the E48 and E96 series, though it is often used on the E12 and E24 series as well. The first three bands indicate the first three digits of the value, and the multiplier is indicated by the fourth band. The fifth band indicates tolerance.

Connected resistances

When two resistances are connected as in Figure 3.2a, we say they are **in series**. To find their total resistance, we simply add them together. For example, if R_1 is 47 Ω and R_2 is 82 Ω, the resistance of the two in series is 47 + 82 = 129 Ω. The same rule applies to three or more resistances, all in series.

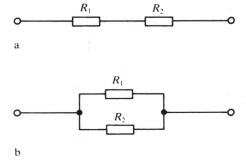

Figure 3.2 *Connecting two resistors (a) in series (b) in parallel*

When two resistances are connected as in Figure 3.2b, we say they are **in parallel**. To find their total resistance R we use this formula:

$$\frac{1}{R} = \frac{1}{R_1} + \frac{1}{R_2}$$

The formula can be extended by adding $\frac{1}{R_3}$ and even more terms to cater for three or more resistances in parallel.

If there are only two resistors in parallel, the formula is simplified to:

$$R = \frac{R_1 \times R_2}{R_1 + R_2}$$

For example, if R_1 is 47 Ω and R_2 is 82 Ω the resistance of the two in parallel is:

$$R = \frac{47 \times 82}{47 + 82} = \frac{3854}{129} = 29.9\,\Omega$$

Note that the resistance of two or more parallel resistances is always less than that of the smallest resistance.

Variable resistors

The resistors described above each have a fixed value. We call them **fixed resistors**. Quite often a circuit needs a resistor which can be varied in value. A typical example is when we want to be able to control the loudness of the sound coming from a radio set. Variable resistors have a track made of resistive material (such as carbon) with a **wiper** which is in contact with the track and moved along it to vary resistance. The support for the wiper is not shown in Figure 3.3a. The resistance between terminal A and the wiper increases as the wiper is moved to the right. At the same time, the resistance between the wiper and terminal B decreases.

Slider resistors or **slider potentiometers**, very similar in appearance to Figure 3.3a are often used as volume controls in audio systems. Five (possibly more) sliders mounted side-by-side allows the response to be adjusted independently for treble, middle and base frequencies. In a **rotary potentiometer** (Figure 3.3b) the track is a 270° sector of a circle and the wiper is mounted on the spindle. The whole is enclosed in a metal or plastic case with three terminal pins, so the track is not visible in the figure. These potentiometers or 'pots' as they are often called have a wide range of uses for controlling electronic circuits. Volume controls, tone controls, brightness controls, speed controls, may all be effected by a 'pot'. A smaller version is the **preset potentiometer**, or **trimpot**, which is not usually enclosed in a case. Its rotary wiper has a slot to take a screwdriver. Trimpots, as their name implies, are intended to be adjusted when a circuit is being set up, or from time to time, to trim the resistance to the best value once and for all.

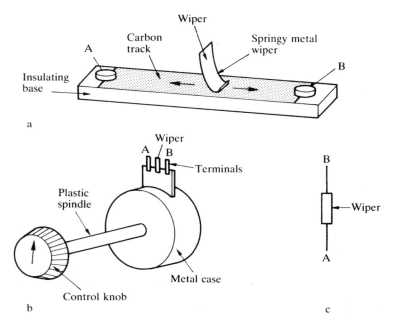

Figure 3.3 *Variable resistors (a) In principle (b) A typical rotary potentiometer (c) Symbol for a variable resistor*

For greater stability, better linearity and longer life, the track of a variable resistor may be made of a metallized ceramic (cermet) instead of a carbon-based material.

Power

As mentioned earlier, when a current flows through a resistance, the charge carriers lose potential energy, which is transferred to the material of the conductor, making it warm, possible very hot or even white hot. In an electric motor, most of the energy is used to make the motor rotate, though a little of it appears as heat. In a TV set, the energy is used to produce light and sound, but some heat is produced too. In fact, heat is nearly always produced and usually represents a waste. The rate at which electrical energy is being used by a device can be calculated very easily, simply by multiplying the current by the p.d. across the device. The result is the **power** of the device, the rate at which it is converting energy from one form (electrical energy) into another form (heat, light, sound, motion). The unit of power measurement is the **watt**, symbol W.

For example, the current through a 6 V torch bulb is found to be 0.5 A. The power of the bulb is $P = 6 \times 0.5 = 3$ W.

This calculation can also be applied to devices which generate current. For example, a solar panel produces a current of 0.8 A and the p.d. across its terminals is 12 V. Its power is $P = 0.8 \times 12 = 9.6$ W.

Superconduction

Certain materials, called **superconductors**, have the ability to conduct electricity without offering any resistance. This phenomenon was first discovered in 1911 by the Dutch physicist, Heike Kemerlingh Onnes. He used a wire of solid mercury and found that it had no resistance when the temperature was reduced below 4 K ($-269°$ C). Kemerlingh passed current into a ring of wire in a liquid helium bath and found that the current was still circulating one year later. The temperature below which a material becomes superconducting is known as the **critical temperature**. Several other substances have since been found to be superconductive, including tungsten, aluminium and lead, but only at critical temperatures of a few kelvin.

Materials of zero resistance have many practical applications including power lines and high-powered electromagnets (see p. 47). Because there is no resistance, no heat is generated and no energy is wasted. Superconducting magnets are used to levitate vehicles above a metal track so that its movement is practically frictionless. In Japan a MAGLEV train hovers in the air above its track when its superconducting magnets are energized. In Germany another MAGLEV train running on a 31.5 km long track has achieved a world record speed of 435 km/h. Another application of superconducting magnets is in the storage of energy. Energy may be stored by building up a high-intensity magnetic field, then released later as current is drawn from the magnet and the field collapses. With the very high field strengths that are attainable with super-conducting magnets, it is possible to store large quantities of energy in a relatively small volume. Storage is 100% efficient because there are no heat losses and all the stored energy is recoverable.

Conversely, a superconducting quantum interference device (SQUID) consists of a superconducting coil which, owing to its zero resistance, is able to produce an appreciable induced current when placed in a very weak magnetic field. SQUIDs have many applications for detecting weak fields or the fields produced by weak currents. For example, a SQUID can detect the electrical field generated by the human heartbeat, and could detect at a distance of 10 km an alternating current amplitude of 1 A flowing in a straight wire.

While superconductivity has many applications, real and potential, there is the disadvantage that superconduction occurs only at low temperatures. This means that superconducting equipment must be cooled, usually by circulating liquid helium around or through it. This adds to the bulk, power requirements, the cost of the equipment and the cost of operating it. Much research time is devoted to finding materials which do not have such low critical temperatures. Year by year new materials are found with higher critical temperatures. The most recently researched materials are ceramics containing substances such as copper oxides, mercury and barium. These have the advantage that they can be formed into wires and sheet and also sprayed on to surfaces of any shape. One such material, discovered by Dr Paul Chu and his associates at the Texas Centre for Superconductivity, University of Houston, has a critical temperature of 153 K. This is still a very low temperature by everyday standards but it does at least allow the superconducting coils to be cooled by liquid nitrogen, which is far cheaper than liquid helium.

Internal resistance

When a cell is not connected into a circuit, there is an e.m.f. between its terminals. The size of the e.m.f. depends on which two metals are used for the electrodes. The e.m.f. causes a p.d. between the terminals. The size of the p.d. equals the size of the e.m.f.

In Figure 3.4a a cell is causing current (conventional current, see p. 12) to flow round a circuit which consists of a resistor, value R. But, to complete this circuit, there are ions carrying electric charge through the electrolyte of the cell.

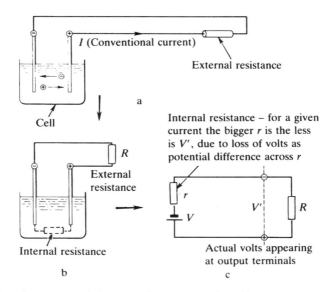

Figure 3.4 *The concept of the internal resistance of a cell*

Like all conductors, the electrolyte has resistance. Taking this into account, we recognize that the circuit has a second resistance, termed the **internal resistance** of the cell (Figure 3.4b). The value r of the internal resistance depends on the composition of the electrolyte, the distance apart of the electrodes and their area and may differ from cell to cell.

Figure 3.4c is a schematic diagram of the circuit. Everything to the left of the dashed line is part of the cell. The circles marked ' + ' and '−' represent its terminals. V represents the e.m.f. of the cell and r is its internal resistance; the diagram represents these two things separately. The external resistor R is to the right of the dashed line. Suppose that a current I is flowing in the circuit, flowing in a clockwise direction. The equation on p. 20 tells us that the p.d. across the internal resistance is Ir. There is a fall of potential as the current flows through the external resistance. This means that V' the p.d. between the terminals must be less than V, the e.m.f. In short, when a cell is connected into a circuit, the p.d. between its terminals is always less than its e.m.f. This is the 'lost p.d.' across the internal resistance.

The 'lost p.d.' is given above as Ir. If I or r or both are very small, the 'lost p.d.' is small and the p.d. across the terminals is almost equal to the e.m.f. of the cell. If the circuit draws only a few milliamps, p.d. and e.m.f. are almost equal, but the more current taken the bigger the difference and the lower the p.d. becomes. This explains why we can not use a battery of zinc–carbon cells to power the self-starter motor of a car. Zinc–carbon cells have relatively high internal resistance. If they are supplying current to a portable radio set, which takes only a few hundred milliamps the p.d. remains almost as high as the e.m.f. But, if we try to run the starter motor from such a battery, the large internal resistance of the cells prevents the battery from supplying enough current (several amps) to turn the motor. Most of the e.m.f. of the battery is dropped as a p.d. across its internal resistance, with practically no p.d. across the motor. The battery may get hot but the motor does not turn. By contrast, the internal resistance of a lead-acid car battery is very low, so it can deliver a large current with relatively little drop in p.d. It also explains why we can obtain an unpleasant electric shock from a 6 V car battery but not from a 6 V torch battery.

4

Capacitance

One commonly used electronic component is the **capacitor**. Its basic structure is very simple (Figure 4.1). It is like a cheese sandwich, consisting of two metal plates (the bread of the sandwich) with a layer of non-conductor (the cheese) between them. The plates have two wires connected to them so that they can be joined to other components. The material between the plates is called the **dielectric**. Some types of capacitor have air as the dielectric but more often the dielectric is a layer of plastic.

Imagine a capacitor connected across the terminals of a cell (Figure 4.2a). Electrons flow from the negative terminal of the cell and enter the plate connected to that terminal. The negative charge on that plate repels the free electrons that are in the metal of the opposite plate. These are the electrons which belong to the outer orbits of the metal atoms but which tend to wander around inside the lattice (p. 7). The electrons flow away, attracted toward the positive terminal of the cell. The atoms they leave behind are now positively charged. This flow of electrons into one plate and out of the other continues until the potential difference (p.d.) between the plates is equal to the p.d. of the cell. Note that, although electrons flow into one plate and out of the other plate, there is no flow of electrons from one plate to the other. Such a flow

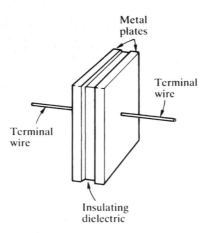

Figure 4.1 *The structure of a capacitor*

30

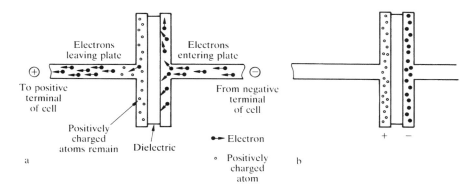

Figure 4.2 *The action of a capacitor (a) Charging (b) Charged*

would be impossible, because the plates are not in contact with each other and the dielectric is a non-conductor.

Suppose that we now disconnect the capacitor from the cell. The plates remain charged as before, and there is still a p.d. between them, equal to the original p.d. of the cell. This illustrates one of the uses of capacitors; they store electric charge (see p. 35). The electrons on one plate remain attracted to the positively charged atoms of the other plate. Once a capacitor has been charged, it holds its charge for a very long time, for hours or even for months. In time, it will become discharged, perhaps by slight leakage of current through the dielectric (it might not be a perfect non-conductor), or when charged ions in the air come into contact with the terminal wires.

Capacitance

The amount of charge a capacitor holds depends on the p.d. between its plates and on its capacitance. Let us consider the p.d. first. If a capacitor holds a certain amount of charge when the p.d. is 1 V, it holds double that amount when the p.d. is 2 V. It holds a hundred times that amount when the p.d. is 100 V. There is an upper limit to how much the charge can be increased by charging it to a high p.d. This is reached when the p.d. is so strong that the electric field between the plates causes the dielectric to break down. Sparks pass through the dielectric, destroying the capacitor. But most capacitors can be charged to a p.d. of 100 V and some can withstand 1000 V or more.

Capacitance can best be understood by comparing capacitors with tanks of water. Both tanks in Figure 4.3 are filled to the same level but the wider one holds much more water than the other. Similarly two different capacitors may be charged to the same p.d. (equivalent to the water level), but the one with

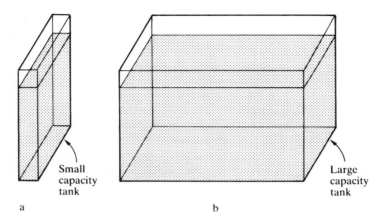

a   b

Figure 4.3   *A tank of water as an analogy of a charged capacitor*

greater capacitance (equivalent to the wider tank) holds more charge (water). We can express these thoughts in a simple equation. Since the amount of charge is proportional to p.d. and to capacitance, the equation is:

$$Q = VC$$

In this equation, Q is the charge, in coulombs (p. 16), V is the p.d., in volts and C is the capacitance, in farads. For example, if a capacitor has a capacitance of 2 farads and is charged until the p.d. between its plates is 5 V, then the amount of charge it is holding is $Q = 5 \times 2 = 10$ coulombs. This is rather a lot of charge; charging such a capacitor to that level requires a current of 1 A to flow for 10 seconds.

Submultiples of the farad

The farad (symbol, F) is too large a unit for practical use. More often we find capacitances quoted in microfarads (symbol μF) where 1 F equals a million microfarads. We also use even smaller units:

1 μF = 1000 nanofarads (symbol, nF)
1 nF = 1000 picofarads (symbol pF)

Practical capacitors

The capacitance of a capacitor depends on four main factors:

- The area of the plates: the larger the area, the greater the capacitance
- The distance between the plates: the closer the plates, the greater the capacitance
- The overlap of the plates: the more they overlap, the greater the capacitance
- The nature of the dielectric.

Figure 4.4 shows the structure of some commonly used types of capacitor. Figure 4.4a is a variable capacitor, often used in tuning circuits in radio sets (p. 155). The dielectric is air. There are two sets of plates, one fixed and the other mounted on the rotating spindle. The fixed plates are electrically connected to each other and they alternate with the set of movable plates which are also connected to each other. The amount of overlap between the two sets, and therefore the capacitance, is adjusted by turning the spindle. The shape of the plates is such that equal turns of the spindle result in equal changes in capacitance. With all except the largest capacitors of this type, the capacitance is only a few hundred picofarads. Part of the reason for this is that the plates are relatively far apart. This is necessary because the plates may become slightly bent in use, with a risk of touching.

Many capacitors have the structure shown in Figure 4.4b, in which the plates are made from thin metal foil separated by sheets of plastic film. The plastic sheet is thin, so bringing the plates close together and increasing the capacitance. The plates have a large area but are rolled together to produce a compact device with relatively high capacitance. The whole is encapsulated, usually by dipping it in fluid plastic, which later hardens. There are many variations on this theme. In some types the plates are made by depositing a metal film on a thin plastic foil; the plastic acts as the dielectric. Capacitors are also made from thin sheets of mica, with silvered surfaces; these have low capacitance but very high stability and are suitable for precision oscillators and tuning circuits. Many different kinds of dielectric are used, conferring special features on the capacitor, such as the ability to withstand high voltage, high capacitance in small volume, high stability in changing temperatures, or suitability for high-frequency operation.

For large values of capacitance, the most frequently used type is the aluminium electrolytic capacitor (Figure 4.4c). The plates are made of aluminium and are rolled together for compactness. The plates are separated by thin sheets of paper soaked in an electrolyte. The main function of the paper is to separate the plates, not to act as a dielectric. In fact, because of the electrolyte a current can readily pass through the paper. The dielectric is formed by applying a p.d. to the plates. This causes a very thin layer of aluminium oxide to form on the anode plate. This layer is non-conducting and acts as the dielectric. The soaked paper, which is conductive, is simply part of the cathode plate. The oxide

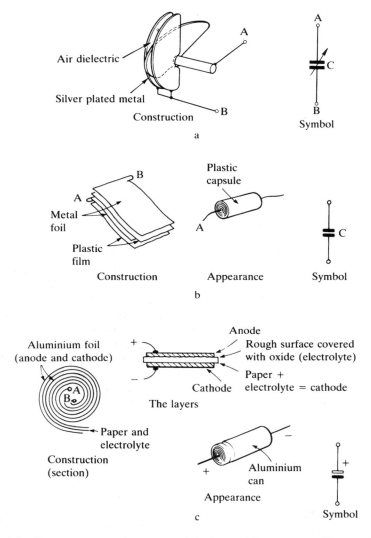

Figure 4.4 *Common types of capacitor (a) A variable capacitor (b) A polystyrene capacitor (c) An aluminium electrolytic capacitor*

dielectric layer is exceedingly thin, so the plates are, in effect, exceedingly close together and capacitance is high. Before the anode is oxidized its surface is roughened, which increases its surface area and makes capacitance even higher. This combination of close spacing and large area makes it possible to achieve capacitances of tens of thousands of microfarads in a capacitor of convenient dimensions. Capacitors measuring only 8 mm in diameter and 21.5 mm long can be made with capacitance as high as 1 F. These are used for storing electric charge as the back-up power supply for the memory of a computer, in case of interruption of the mains power supply. GoldCap double-layer capacitors have capacitance values measured in farads and are used for similar purposes. They may also be used instead of ordinary batteries in bicycle lamps and electric shavers. One of their advantages compared with rechargeable cells is that they can be recharged rapidly and no special charger is required. Another advantage is that capacitors may be recharged thousands of times without any deterioration in their performance. For the viewpoint of the conservationist, capacitors are preferable to cells because they do not contain toxic heavy metals such as lead or mercury.

Danger from capacitors

A disconnected charged capacitor holds its charge for a long time. For this reason, a capacitor may retain its charge long after a piece of equipment has been switched off. It is able to deliver the charge very quickly when you touch its terminal wires, producing a dangerously, possibly fatally, high current. Even if a capacitor has apparently been discharged by briefly touching its terminal wires together, it may still retain some of its charge in the dielectric. This residual charge is later transferred to the plates, and could still deliver a powerful electric shock. For this reason, a large-capacitance capacitor should always be stored with its terminal wires twisted together, to ensure that it is fully discharged.

Aluminium electrolytic capacitors have wide tolerance, typically $\pm 20\%$, but sometimes greater. Their capacitance falls if they have not been in use for several weeks, but is gradually restored by subsequent use. The lack of precision and stability makes electrolytic capacitors unsuitable for tuned or timing circuits. Their main disadvantage is that they are polarized, which again limits the kinds of circuit in which they can be used, They must always be connected with the anode positive of the cathode, otherwise the oxide layer is eventually destroyed and the capacitance is lost. The case is marked to show the correct polarity. Leakage current in electrolytic capacitors is appreciably higher than that of capacitors of other types.

Another type of electrolytic capacitor consists of a sintered block containing particles of tantalum. For their size, these tantalum bead capacitors have a high capacitance but their breakdown voltages are low and, like the aluminium type, they are damaged by reverse p.d.s. Modern sub-miniature aluminium electrolytic capacitors have almost the same high capacity in small volume as the tantalum beads.

Charging a capacitor

Let us look more closely at what happens when a capacitor is charged. In Figure 4.5 a capacitor C is fed from a constant source of p.d. (or constant voltage source) through a resistor. The source might be a 6 V battery. At every stage in what follows, the p.d. across the resistor is given by:

p.d. across resistor = p.d. of source − p.d. across capacitor   (1)

Another equation that holds true at every stage is derived from the equation on p. 20:

$$\text{current through resistor} = \frac{\text{p.d. across resistor}}{\text{resistance}} \qquad (2)$$

Figure 4.6a shows what happens to the capacitor p.d. The capacitor is uncharged to begin with so the p.d. is zero. Current flows through the resistor and into the capacitor; charge builds up rapidly and the p.d. begins to rise. But, as the p.d. starts to increase, and because the p.d. of the source is constant, the p.d. across the resistor is reduced (equation (1)). This reduces the current through the resistor (equation (2)). As a result, current flow into the capacitor is reduced, charge does not accumulate as quickly and p.d. rises less rapidly.

Figure 4.6a shows the p.d. rising rapidly at the start but gradually levelling out as time passes. Eventually there is no further increase. This is when the p.d. has become equal to that of the source. Now there is zero p.d. across the resistor (equation (1)). There is no current through the resistor (equation (2)) and charging ceases.

The curve of Figure 4.6a has a definite shape (we say that it is exponential) but we do not need to delve into its mathematics. There is a useful way of specifying a key feature of the curve. This is the **time constant**, which is defined as the time taken for the p.d. to rise to 63% of the source p.d. If t is this time, in seconds, R is in ohms and C in farads, it can be shown that:

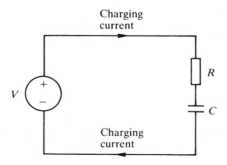

Figure 4.5   *Charging a capacitor*

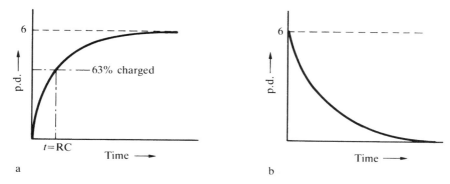

Figure 4.6 *Graphs of the p.d. across a capacitor against time when being (a) charged (b) discharged*

$$t = RC$$

For example, if the resistance is 3.3 kΩ and the capacitance is 10 μF, the time constant is 3300 × 0.000 01 = 0.033 s. It takes 0.033 s for the p.d. to rise to 63% of the source p.d.

Time constant

The larger the resistance and the larger the capacitance, the longer the time constant.

A capacitor is allowed to become fully charged (by which we mean to have charged to the same p.d. as the source), and is then discharged by replacing the source by a wire (Figure 4.7). The p.d. falls rapidly at first (Figure 4.6b). As the capacitor becomes partly discharged, the p.d. across the resistor is reduced and discharge is slower. The p.d. falls more and more slowly. In theory it never reaches zero, but in practice it reaches zero in approximately 3 time constants. In the example above, the capacitor would take about 3 × 0.033 = 0.1 s to discharge practically to zero.

Alternating current

Some oscillator circuits, such as the one described on p. 156, produce a p.d. which continually reverses in direction. A graph of the p.d. plotted against time (thick continuous line, Figure 4.8) shows the p.d. starting at zero, then increasing to a maximum value in the positive direction. Having reached its maximum

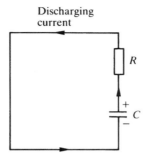

Discharging
current

R

+
C
−

Figure 4.7   *Discharging a capacitor*

it decreases to zero again, then begins to increase in the negative direction. After reaching a maximum negative value it decreases to zero again. This completes one cycle, which is then repeated indefinitely. The number of cycles completed in one second is the **frequency** of the oscillator, in cycles per second. Another term for cycles per second is **hertz**, symbol Hz, which is the unit of frequency.

The exact shape of the p.d.–time curve depends on the nature of the oscillator. The oscillator on p. 156 produces a repeating shape (or waveform) which is identical with the curve obtained when we plot a graph of the sine of angles from 0° to 360° (that is, for one cycle, the figure shows two cycles). We say that it is a sine wave oscillator. If we connect a resistor across the output terminals of this oscillator the flow of current at any given time is proportional to the p.d., since current = p.d./resistance (p. 20). The graph of current against time has the same shape as Figure 4.8. The current flows in one direction then reverses and flows in the other direction, then reverses again, repeating the reversals indefinitely. We say that it is an **alternating current**. This is sometimes abbreviated to a.c. This contrasts with the current obtained when a resistor is connected across a cell. Then we obtain a steady current flowing in one direction only, which is known as **direct current**, often shortened to d.c.

It is interesting to see what happens if we connect a resistor and capacitor in series across a sine wave oscillator (Figure 4.9). Figure 4.10 shows how the p.d.s across the oscillator, the resistor and the capacitor vary during the course of three cycles. It also shows the current through the resistor. The graph of the source p.d. is that of a sine wave for which the maximum p.d. in either direction is 1 V. We say its amplitude is 1 V. Its frequency is 1000 cycles per second, or 1000 Hz, better expressed as 1 kilohertz (1 kHz). The resistor is 330 Ω and the capacitor is 0.5 μF.

The graph is plotted with milliseconds on the *x*-axis and millivolts on the *y*-axis. It shows the p.d. across the source (continuous thick line) oscillating as a sine wave between +1 V (+1000 mV) and −1 V (−1000 mV) for 3 milliseconds, the time required for 3 cycles at 1000 cycles per second.

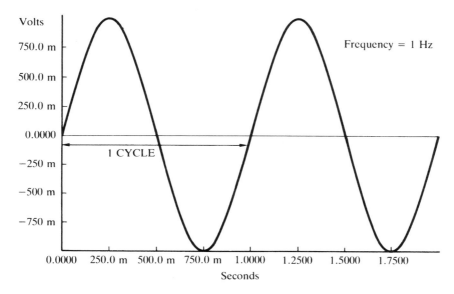

Figure 4.8 *Sine waves, as produced by a sine-wave oscillator*

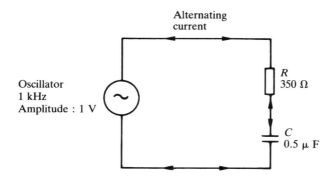

Figure 4.9 *Charging and discharging a capacitor with a sine-wave oscillator*

Now look at the continuous thin-lined curve, the resistor p.d. This too is a sine wave but it does not reach such high values in either direction as the source p.d. This is understandable because the total p.d. from the source is split into two parts, the resistor p.d. and the capacitor p.d. At any instant of time, the resistor and capacitor p.d.s must add up to the source p.d. If you check the graphs, you can see that this is so.

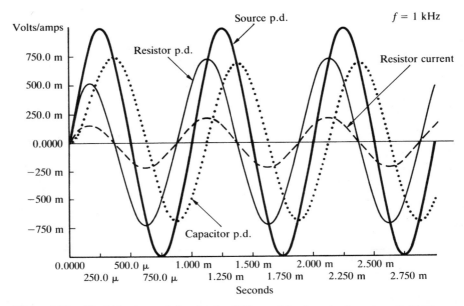

Figure 4.10 *The behaviour of the circuit of Figure 4.9 when the frequency is 1 kHz*

The alternating changes of p.d. across the resistor and capacitor and the alternating current through the resistor follow a complicated but repeating pattern. There are certain important points about this pattern:

- The waveform of all four quantities is a sine wave.
- All four waveforms have the same frequency.
- The resistor p.d. reaches its peaks in either direction slightly earlier than the source p.d.
- The resistor current (dashed line) reaches its peaks at the same time as the resistor p.d. This current, incidentally, is plotted on a scale 100 times larger than that used for the p.d.s, so that the maximum current shown in Figure 4.10 is 1.5 mA.
- The capacitor p.d. reaches its peaks in either direction slightly later than the source p.d.
- The amplitude of the capacitor p.d. is only slightly less than that of the resistor p.d.

The fact that the p.d. sine waves reach their peaks at different times is expressed by saying that they are out of phase, or that there are phase differences between them. The usual way of expressing a phase difference is to quote the phase angle. In the graph, the resistor p.d. reaches its peak about 0.128 ms before the source p.d. in each cycle. If 1 ms (the time of one cycle) is equivalent to 360°, then 0.128 ms is equivalent to $360 \times 0.128 = 46°$. We say that the resistor p.d. has a phase angle of $+46°$, or we can say that it has a phase lead of 46°. The resistor current is in phase with the resistor p.d. (This is inevitable; Ohm's Law applies at every instant, so the p.d. and current of a resistor are always in phase.)

Similarly, the capacitor p.d. reaches its peak about 0.013 ms after the source p.d. This is equivalent to a phase angle of

$$-(360 \times 0.135) = -49°.$$

This is a phase lag of 49°.

Before we leave Figure 4.10, note that the curves during the first cycle do not have quite the same shape as during the second and subsequent cycles. This is because the capacitor has no charge on it at the beginning of the first cycle. From the second cycle onwards, it already has charge left over from the end of the previous cycle and the waveform is identically shaped in each cycle.

Effects of frequency

We have already seen (Figure 4.6) that time is a factor in the charging and discharging of a capacitor. If we increase the frequency of the sine wave oscillator we force the capacitor to charge and discharge in a shorter time, and might expect it to behave differently. Figure 4.11 shows what happens if the frequency is increased tenfold, from 1 kHz to 10 kHz. As before, we show three cycles, but now they take only 300 μs. The same curves are plotted but the pattern they make is very different. The most noticeable effect is that the waveform of the resistor p.d. is almost the same as that of the source p.d. It has almost the same amplitude and only a very small phase lead. By contrast, the capacitor p.d. has a much smaller amplitude than before, and its phase lag has increased to about 90°.

Most of the source p.d. now appears across the resistor, and only a small fraction of it appears across the capacitor.

This result is almost what we would expect if we replaced the capacitor with a resistor of very low value. Because of Ohm's Law, there could be only a small p.d. across it. The capacitor is acting like a low-value resistance.

If we shift the frequency in the opposite direction, making it ten times less than in Figure 4.10, we obtain Figure 4.12, in which the frequency is 100 Hz. Now it is the source p.d. and the capacitor p.d. which are alike. The amplitude of the capacitor p.d. has increased and its phase lag has almost disappeared. By contrast, the resistor p.d. has reduced in amplitude and its phase lead increased to about 90°. At this low frequency, the effect of the resistor is much less significant than that of the capacitor. It is as if the capacitor has been replaced by a resistor of very high value, so that currents create a large p.d. across it. The capacitor is acting like a high-value resistance.

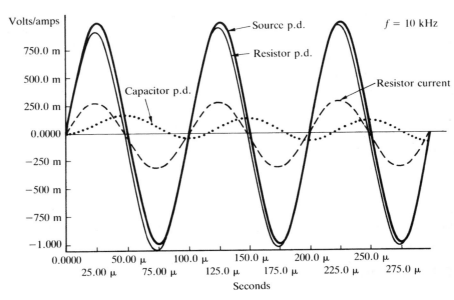

Figure 4.11   *The effect of increasing the frequency to 10 kHz*

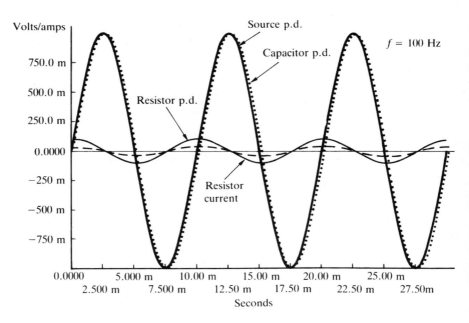

Figure 4.12   *The effect of reducing the frequency to 100 Hz*

Summing up:

- At high frequencies a capacitor acts like a low-value resistance; high-frequency signals pass easily through it
- At low frequencies it acts like a high-value resistance; it blocks the passage of low-frequency signals.

We return to this topic in Chapter 15.

Stray capacitance

Capacitance in a circuit is not limited to that provided by its capacitors. Many other components have capacitance because of their structure (p. 73). Stray capacitance is also produced when two conducting tracks on a circuit board are very close together, acting as capacitor plates. Such capacitance usually amounts only to a few picofarads and can be ignored. But in high-frequency circuits, such as microwave circuits (p. 257), the effects of capacitance are much greater and stray capacitance must be allowed for.

Coupling

The dielectric in a capacitor is a non-conductor, so no actual current can pass through a capacitor. But in Figure 4.13 an alternating signal is able to pass freely through a capacitor from circuit A (which might be a microphone) to circuit B (which might be an amplifier). The capacitor is coupling the two circuits together.

The circuit on the left of Figure 4.13 is connected to plate A of the capacitor. Assuming that this is a microphone, it generates an alternating potential (though not necessarily a regular sine wave) when it detects sound. When there is no sound the potential of plate A is zero. When there is sound the

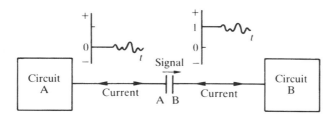

Figure 4.13   *A coupling capacitor*

potential alternates about zero, sometimes being positive and sometimes being negative. If the circuit on the right is an amplifier, the potential at plate B will need to be about 1 V in order for the amplifier to work properly. This potential is provided by biasing resistors (p. 169) not shown in the figure.

During periods of silence, there is an unchanging p.d. of 1 V between A and B. No current is flowing on either side of the capacitor. When sound is detected, the alternating signal from the microphone causes current to flow into and out of plate A. The potential at plate A oscillates above and below zero. But, as we have seen above, a capacitor acts as a low resistance to high-frequency signals. The signal passes through the capacitor so that currents flow in and out of plate B. Its potential oscillates above and below its constant potential of 1 V. The difference between the potential levels at A and B are shown by the small graphs in Figure 4.13.

In general, capacitors may be used to pass signals from one part of a circuit to another part when the average potentials of the two parts differ.

Decoupling

In Figure 4.14 there are two circuits A and B, which take their current from the same power source. The power requirements of circuit A are liable to change very rapidly. Any sudden change in the amount of current drawn from the power line causes the p.d. between the lines to rise and fall suddenly. Sudden rises and falls in its power supply can upset the action of circuit B. One way to avoid this is to connect a capacitor between the power lines. We use it to isolate circuit B from the effects of circuit A. In other words, it decouples A and B.

Decoupling works like this. A sudden change in the p.d. between the lines is the equivalent of a high-frequency signal. Although it may be only one pulse of

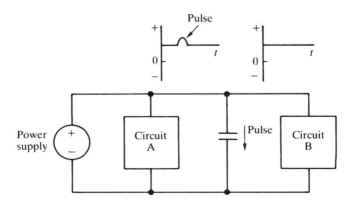

Figure 4.14   *A decoupling capacitor*

higher or lower p.d. (equivalent to only half a cycle of a.c.), it is a signal that is easily passed through the capacitor. The pulse on the power lines in the region of circuit A passes through the capacitor from one power line to the other. The capacitor prevents it from reaching circuit B.

Capacitors for batteries

Developments in the technology of capacitor construction mean that new high-capacity capacitors may well replace chemical cells in many applications. New types of capacitor known as PowerCaps have capacities up to 1.5 kilofarads at 3 volts. A simple calculation shows that the stored charge is 3 × 1500 = 4500 coulombs. This means that such a capacitor can deliver an average current of 1 A for 4500 seconds, or one and a quarter hours. An application of interest to automotive engineers is to use the capacitors to deliver very large currents (several hundred amps) for relatively short periods. In this way PowerCaps can supplement the power supply from the vehicle's conventional lead-acid battery at moments when it is overloaded. In the future such capacitors may replace the lead-acid battery completely.

Uses of capacitors

Summing up the discussions of this chapter, capacitors are used for:

- Storing electric charge
- Setting the time constant of timing and oscillating circuits
- Coupling; transferring alternating signals from one circuit to another
- Decoupling; preventing disturbances on the power lines from spreading from one circuit to another.

There is a little more about capacitors on pp. 61–3.

5

Inductance

An inductor consists of a coil of wire, usually surrounding a core of ferromagnetic material. The ferromagnetic material may be iron itself or an iron-containing material known as **ferrite**. Figure 5.1 shows an inductor in which the core is made from layers of sheet iron. Another kind of inductor, is the aerial of a portable radio set, which consists of a rod of ferrite on which is wound one or more tuning coils.

Electricity and magnetism

When a current flows along a straight wire, a circular magnetic field is generated around the wire, as shown in Figure 5.2. In the diagram, the needle of a small compass is placed close beside the wire points in the direction of the field. If we reverse the direction of the current the direction of the field is also reversed. If the wire is made into a coil, as in Figure 5.3 the fields due to the individual turns of the coil are combined to produce a field in the coil. With a core present as shown, the field is mainly confined to the core, and is stronger.

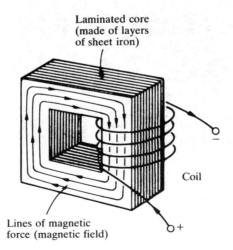

Figure 5.1 *An iron-core inductor; only a few turns of the coil are shown*

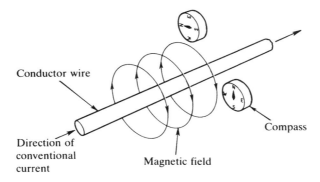

Figure 5.2 *The magnetic field produced by a current flowing through a wire*

The field is made stronger still if there are many turns in the coil and if the current is large. A coil with a core inside it to concentrate the field may have practical uses as an electromagnet. Large electromagnets are capable of lifting heavy loads, such as metal car bodies. We mention some other uses of electromagnets later.

Magnetic poles

When the current flows in the direction shown in Figure 5.3 the lines of magnetic force enter the coil on the right and leave it on the left. The coil is equivalent to a magnet with its south-seeking pole to the right and its north-seeking pole to the left.

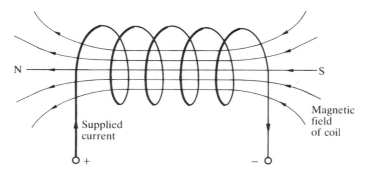

Figure 5.3 *The magnetic field produced by a current flowing through a coil*

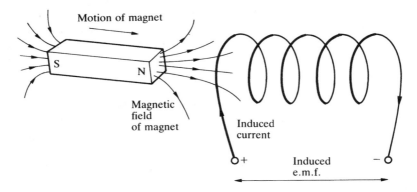

Figure 5.4   *A current induced by moving a magnet toward a coil*

In the electromagnet, the current generates a magnetic field. The reverse effect is found when we take a coil and introduce a magnetic field into it (Figure 5.4). As the magnet is pushed toward one end of the coil and the field of the magnet begins to enter the coil, a current is generated in the coil. We say that the magnetic field has **induced** a current in the coil. When the magnet is pulled away the induced current reverses in direction. The most interesting point about induction is that the current is produced only if the magnetic field is changing. A current flows while the magnet approaches and enters the coil, but ceases to flow as soon as the magnet stops moving. The amount of current induced depends on the rate of change of the magnetic field. If we thrust the magnet into the coil rapidly, the current is large. If we move it in slowly, the current is small. Of course, the overall effect is the same, with slow movement a smaller current flows, but for a longer time.

Another key fact about induction is that the direction of the current is such as to oppose the movement of the magnet. This effect is known as Lenz's Law. Comparing Figure 5.3 with Figure 5.4 we see that the induced current in Figure 5.4 flows in the same direction as the supplied current of Figure 5.3. The magnet has its north pole toward the coil as it approaches. The field produced by the induced current is in the opposite direction, with the north pole toward the magnet. The two like poles repel each other. We have to do extra work to overcome this repulsion as we push the magnet into the coil. This extra work provides the energy which generates the current.

Self induction

With the above facts of electromagnetism and induction in mind, look again at Figure 5.3. When there is no current supplied to the coil there is no magnetic field. Now suppose that the current is switched on. Immediately the coil acts as an electromagnet. The effect of this is the same as if we had suddenly placed a magnet inside the coil. A current is induced in the coil to oppose the field already present. This induced current is the result of the field that the coil has itself generated. This effect is known as **self induction.**

The result of self induction is that any change in the amount of current supplied to the coil (turning it on, turning it off, increasing it, or decreasing it) is opposed by a self-induced current. If we increase the supplied current, the induced current acts to oppose the increase, to hold the current constant. If we reduce the supplied current, the induced current is in the same direction as the supplied current, acting so as to oppose the decrease, to hold the current constant.

Inductors and inductance

The extent to which an inductor opposes changes in the current passing through it is known as its self-inductance. The unit of self-inductance is the **henry**, symbol H. An inductor of 1 H needs many turns of wire wound on a massive core. Most practical inductors have inductances rated in millihenries (symbol, mH) or microhenries (symbol, µH).

This effect is easily demonstrated in a circuit such as Figure 5.5. In contrast to a capacitor circuit (Figure 4.9) which is broken by the dielectric, the wire of the inductor coil L allows current to flow through it. In most inductors the wire itself has very low resistance, perhaps only a few ohms, often less than an ohm. The coil resistance is much lower than that of resistor R in Figure 5.5 and may be ignored. When an alternating current amplitude 1 V, frequency 1 kHz is

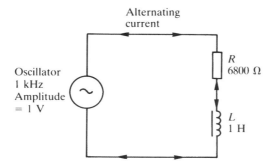

Figure 5.5 *A circuit for passing an alternating current through an inductor*

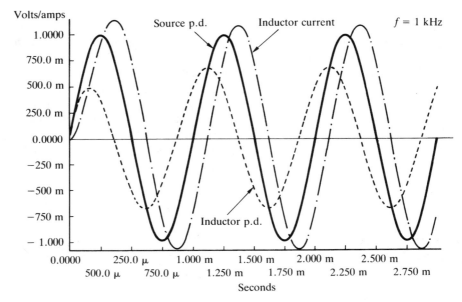

Figure 5.6   *The behaviour of the circuit of Figure 5.5 when the frequency is 1 kHz*

supplied to this circuit the resulting waveforms are as shown in Figure 5.6. This plots the source and inductor p.d.s and the current through the inductor. The inductor current is plotted on a scale that is 10 000 times larger than the scale for p.d.s. We read the maximum current as 108 μA. As with the capacitor circuit of Figure 2.9, we notice certain features of the waveforms:

- The waveform of all three quantities is a sine wave
- All three waveforms have the same frequency
- The inductor p.d. reaches its peaks in either direction slightly earlier than the source p.d.
- The inductor current reaches its peaks in either direction slightly later than the source p.d.

Note that the inductor current and inductor p.d. are not in phase; the inductor is obviously not behaving as a simple resistance (p. 41).

Since the size of an induced current depends on the rate of change of the magnetic field, we expect the effects of self-induction to be greater at higher frequencies. Figure 5.7 shows what happens at 10 kHz. The inductor current is much less than it is at 1 kHz, peaking at only 25 μA, as if the inductor is offering an increased resistance. It is approximately 90° out of phase with the source p.d. The source p.d. and inductor p.d. are approximately equal to one another and almost in phase. At every instant, most of the source p.d. is

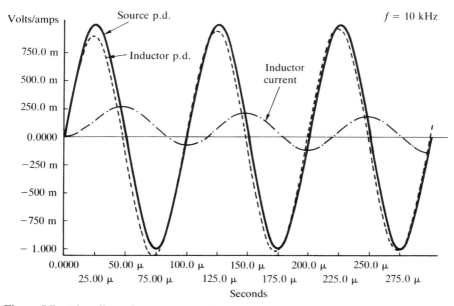

Figure 5.7 *The effect of increasing the frequency to 10 kHz*

appearing across the inductor, leaving only a small fraction of the source p.d. to appear across the resistor. This suggests that the effective resistance of the inductor at high frequency is considerably greater than that of the resistor, perhaps ten times as great.

The pattern of the waveforms changes when the frequency is reduced to 100 Hz. The reduced rate of change of current should mean that the effects of self-inductance are not as obvious. Figure 5.8 shows that this is true. Now it is the inductor current that is in phase with the source p.d. The inductor is behaving as if it were a pure resistance. The current through the inductor peaks at 147 µA, when the source p.d. is 1 V. A calculation of the expected current in the circuit with the 6800 Ω resistor and an inductor coil of negligible resistance we find, using the equation of p. 20, that the expected current is 1/6800 = 147 µA. This shows that the inductor exerts no resistance (or more precisely, a negligible resistance) to the flow of current at 100 Hz. The inductor p.d. leads the source p.d. by about 90°.

Summing up:

- At high frequencies an inductor acts like a high-value resistance; it blocks the passage of high-frequency signals.
- At low frequencies it acts like a low-value resistance; low-frequency signals pass easily through it.

We return to this topic in Chapter 15.

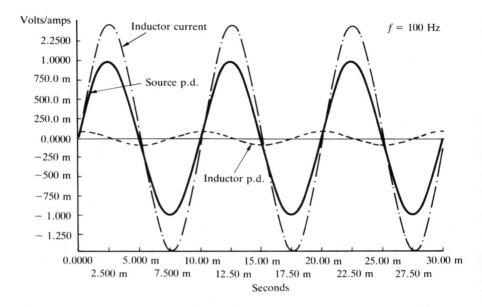

Figure 5.8   *The effect of reducing the frequency to 100 Hz*

Referring back to p. 43, we see that capacitors and inductors offer 'resistance' to the flow of current. Unlike the resistance of a resistor, their 'resistance' depends on frequency. We refer to this as **reactance**, because the effect is produced by the reaction of the capacitor or inductor to the signal being applied to it. Like true resistance, reactance is measured in ohms. But reactance depends on frequency so, when we quote the reactance of a capacitor or inductor, we must state the applicable frequency.

A word that covers resistance and reactance is **impedance**. The impedance of a component or circuit is the sum of its resistance and both kinds of reactance. It is expressed in ohms, at a stated frequency.

High p.d.s in inductive circuits

If a circuit contains an inductor or an inductive component such as a relay (p. 118), it may be subject to unexpectedly high voltages. Suppose that current is flowing through the inductor, then it is switched off. The magnetic field in the inductor collapses immediately. This is a very rapid change and because of this a very large current is induced, such as would be produced by a p.d. of several hundreds of volts. The current flows in the same direction as that in which current was previously flowing, to try to maintain the field. The induced current may be many times greater than originally flowing, perhaps so large that it damages components such as transistors in nearby parts of the circuit. It also causes intense sparking at switch contacts as they are opened. This may damage the contacts and possibly cause them to become fused together. Because of the effects of inductance, special precautions have to be taken in circuits which switch inductive components rapidly (p. 93).

Practical inductors

Figure 5.9 illustrates some types of inductor used in electronic circuits. In Figure 5.9a we see the simplest type of inductance-producing device, ferrite beads. Instead of connecting an inductor into a circuit, we simply thread a few beads on to the conductor which is carrying the high-frequency signal that we wish to suppress. The circular field around the wire (Figure 5.2) creates a circular field in the beads. If this field changes, it induces a current in the conductor to oppose that change. The inductor in Figure 5.9b is similar to the type illustrated in Figure 5.1. The core is like a rectangular figure-8, with the coil wound around the part between the loops of the 8. The coil is of many turns to give high self-inductance. This type of inductor is used to prevent high-frequency signals from passing. It might be used to decouple (p. 44) a part of a circuit that is oscillating at high frequency to prevent the high frequency from reaching other parts of the circuit, where it is unwanted. An inductor used in this way is called a **choke**, because it chokes or blocks high-frequency signals while allowing low-frequency signals to pass. It might often be connected in a power line. A similar choke effect is often obtained by threading ferrite beads on to the power conductors, or winding the power conductors around a ring (a torus) made of ferrite as in Figure 5.9c.

Eddy currents

The magnetic fields in an inductor may generate electric currents in the core. Such eddy currents represent a waste of energy, leading to the generation of heat and interfering with the proper operation of the inductor. One way of reducing eddy currents is to make the core from sheets of iron cut to the same shape. The core is laminated, as in Figure 5.1.

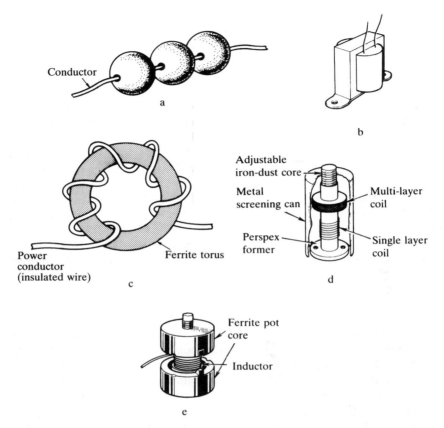

Figure 5.9   *Common types of inductor (a) Ferrite beads (b) Heavy duty choke inductor (c) A ferrite torus (d) Tuning coil (e) Ferrite pot inductor*

A tuning coil is seen in Figure 5.9d. One or more coils are wound on a plastic former. The iron-dust core is threaded, as is the inside of the former; the core is moved into or out of the coil by turning it. This allows the self-inductance of the coil to be adjusted precisely. If there are two or more coils, the core links them electromagnetically, so they function as a transformer (p. 55). Movement of the core adjusts the linking between the coils when setting up the circuit. The inductor is enclosed in an earthed metallic can to prevent magnetic interference between the coils and nearby parts of the circuit.

The inductor shown in Figure 5.9e, known as a pot core consists of a two-part ferrite core with the coil wound around the central column. When the two parts are screwed firmly together the coil is totally enclosed in the core. This type of inductor provides a relatively high inductance in a small volume, and is immune from electromagnetic interference.

Transformers

A transformer is made by winding two coils on a single core. Most often the core is shaped to form a ring, so that all the magnetic field produced by one coil passed through the other coil (Figure 5.10). The coil to which current is supplied is known as the primary coil. When alternating current passes through this it generates an alternating field in the core. The alternating field induces a current in the secondary coil. Note the use of the word 'alternating' in the previous sentence. Current is induced only when a magnetic field is changing (p. 48). A transformer requires alternating current. It does not work with direct current.

The action of a transformer depends on the numbers of turns in the primary and secondary coils. The ratio of the number of secondary turns to the number of primary turns is the **turns ratio**, n, where:

$$n = \frac{\text{secondary turns}}{\text{primary turns}}$$

In a **step-up transformer**, n is greater than 1. When an alternating current of given amplitude passes through the primary coil, the alternating current (at the same frequency) which is induced in the secondary coil has an amplitude n times as great. For example, suppose the primary has 10 turns and the secondary coil has 40 turns. The turns ratio is given by $n = 40/10 = 4$. If the current fed to the primary has an amplitude of 2.5 V, the current in the secondary coil has an amplitude of $4 \times 2.5 = 10$ V. Step-up transformers are used at power

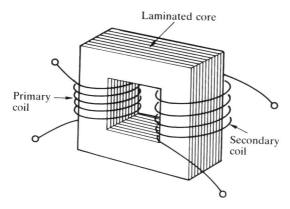

Laminated core

Primary coil

Secondary coil

Figure 5.10   *A transformer*

stations to step up the voltage from the generator to many kilovolts before the power is fed to the distribution network.

If the turns ratio is 1 (primary and secondary turns are equal in number), there is no change in voltage. It might seem that there is little point in having a transformer of this kind. They are used to isolate one circuit from another. For example, we may have equipment which operates at mains voltage (240 V) but for reasons of safety we do not want to have it connected directly to the mains. An isolating transformer with its primary connected to the mains and its secondary connected to the equipment provides the required isolation.

A **step-down transformer** works in the opposite sense to a step-up transformer and has a turns ratio of less than 1.

Such transformers are often in use in mains-powered electronic equipment. Devices such as radio receivers, audio amplifiers, electronic keyboards, and personal computers operate on low voltages such as 12 V, 9 V or less. The first stage in providing a power supply for these is to step-down the voltage from 240 V to the level required. Transformers may also be used for coupling microphones and loudspeakers to amplifier circuits and for coupling different sections of amplifier, radio and oscillator circuits (p. 156)

Power

The primary coil of a transformer converts electrical energy into magnetic energy. Its rate of working or power is given by this equation

$$P = IV$$

This power is recovered in the secondary coil, where magnetic energy is converted back into electrical energy. There can be no gain of power during these processes. The principle of the conservation of energy prevents this. In fact there will be a certain amount of power loss, due to heat produced in the coils and heat produced in the core due to eddy currents (p. 53). Good transformer design minimizes power loss to a small percentage. Assuming that there is no loss (that the transformer is 100% efficient) the value of P must be equal on both sides of the transformer. So, if V is increased four-fold, for example, then I must be reduced to quarter. The maximum current that can be drawn from the secondary coil and is limited to one quarter of that supplied to the primary coil. Usually, the primary coil of a step-up transformer is wound in heavy-duty wire to allow a large current to be supplied, while the secondary coil consists of many turns of fine wire. The reverse applies to a step-down transformer.

6
Simple circuits

Resistors, capacitors and inductors are described as **passive** components. They may have p.d.s applied to them or currents passed through them but they respond in a rather orderly way according to simple rules, such as Ohm's Law, for example. This contrasts with the much more complicated behaviour of the **active** components such as transistors, which we shall examine in Chapters 7 to 10. Before we go further we look at ways in which the passive components are connected to make useful circuit building blocks.

Potential divider

This consists of two or more resistors wired in series. In Figure 6.1a we have a potential divider made up of ten equal resistors, resistance $1\,\Omega$ each. For each resistor, $V = IR$ (p. 20) and, because the same current flows through all of them and they all have the same resistance, I and R are the same for all, making V the same for all. As we go down the chain the potential drops by the same amount across each resistor. From top to bottom, it drops from V to zero in ten equal steps, each of $V/10$.

In Figure 6.1b we have just two resistors, $3\,\Omega$ and $7\,\Omega$. Think of the $3\,\Omega$ resistor as being equivalent to three $1\,\Omega$ resistors in series. It is equivalent to three of the steps of Figure 6.1a. The p.d. across it equals three steps of $V/10$, making a drop of $3V/10$. Similarly, the drop across the 7 W resistor is $7V/10$. If $V = 20$ volts, the drops are $6V$ and $14V$ respectively.

The p.d. across either one of the resistors is proportional to its resistance. In Figure 6.1c we have two resistors totalling $320\,\Omega$. These are the equivalent of 320 individual resistors, each of $1\,\Omega$. There are now 320 equal steps, and the p.d. per step is $V/320$. The p.d.s across the two resistors are:

$$V_1 = 100V/320 \text{ and } V_2 = 220V/320$$

If the value of V is 16 V, the p.d.s across the resistors are $(100 \times 16)/320 = 5$ volts, and $(220 \times 16)/320 = 11$ volts.

Once again, the p.d.s are proportional to the resistances.

A circuit such as this, which divides the total p.d. into two or more parts, is called a **potential divider**. This can be useful in circuits where, for example, the battery provides a p.d. of 6 V but we need a p.d. of 5 V or some other value less

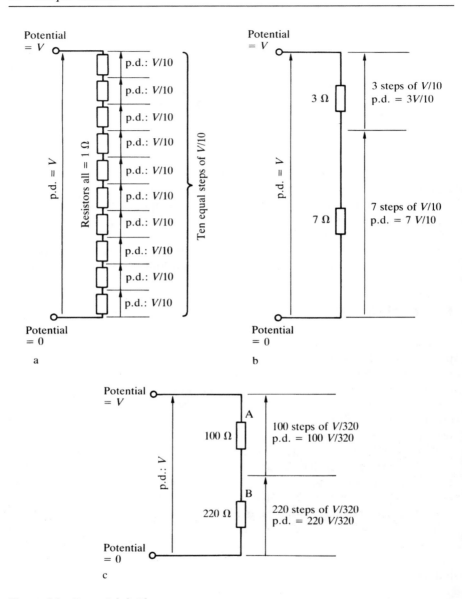

Figure 6.1   *Potential dividers*

than 6 V. A general rule for finding the potential V' at the junction between two resistors A and B (Figure 6.1c) is:

$$V' = \frac{V \times \text{resistance of resistor B}}{\text{total resistance of A and B}}$$

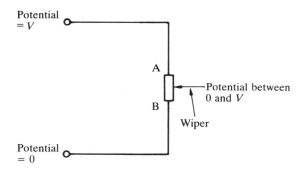

Figure 6.2 *A variable potential divider*

A potential divider can be built from a variable resistor, as in Figure 6.2. Here the two resistors of Figure 6.1b are represented by the two sections of the resistor track, from A to the wiper and from the wiper to B. As the wiper is moved along the track the total resistance (and therefore the number of steps and the p.d. per step) remains unchanged. But as the wiper is moved, the potential at the wiper falls smoothly from *V* when the wiper is at A, to zero when the wiper is at B. This is a way of obtaining any potential within a given range.

The calculations about potential dividers assume that the same current flows along all the resistors. If another circuit is connected to the divider, this may not be true. In Figure 6.3a a circuit with very high resistance is connected to the divider. It takes so little current that almost all of the current flowing through

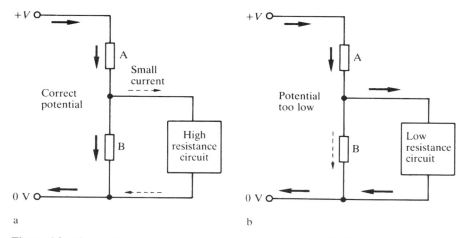

Figure 6.3 *Connecting a low-resistance circuit to a potential divider produces an error*

A goes on through B and the divider gives the calculated potential. In Figure 6.3b the connected circuit has low resistance. Most of the current that flows through A goes on to flow through the circuit, and only a small current is left to flow through B. An unexpectedly small current through B means that the p.d. across it is much lower than calculated. This effect may be important when making measurements with test-meters (p. 161).

Resistor bridge

A bridge is made by connecting four resistors as in Figure 6.4. It can be shown by thinking of R_A and R_B as a potential divider and R_C and R_D as another potential divider, that if:

$$\frac{R_A}{R_B} = \frac{R_C}{R_D}$$

then the potential at X is exactly equal to that at Y. We say the bridge is **balanced**. The milliammeter shows no current flowing between X and Y. The point about using a bridge is that it is very sensitive to being put out of balance. If any one of the values changes even by a small amount, the bridge loses balance and a current flows through the meter.

This type of circuit is used for measuring resistance, with the unknown resistance at R_A, a high-precision variable resistor at R_B and two high-precision fixed resistors at R_C and R_D. R_B is adjusted until the bridge balances (zero current through the meter). Then knowing R_B, R_C, and R_D, the value of R_A is calculated. A bridge is often used with resistive sensors, such as strain gauges, and the idea is adaptable to measuring capacitance and inductance.

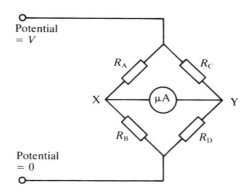

Figure 6.4   *A bridge circuit*

Filters

Figure 6.5a shows a circuit unit that is often used in audio and radio circuits and occasionally in others. If its input V_{IN} is constant, C soon charges to a constant p.d. and the output V_{OUT} settles at a value equal to V_{IN}. The circuit becomes of more use when the input is an alternating p.d., amplitude V_{IN}. To see what happens, re-draw the figure as in Figure 6.5b. It has the same arrangement of parts as a potential divider (Figure 6.1), but with a capacitor replacing the lower resistor. The reactance (equivalent to resistance, p. 52) of the capacitor depends on the frequency of the signal, the alternating p.d. It is high for low frequencies and low for high frequencies.

If the signal has low frequency, the circuit behaves as if it is a potential divider with a very high-value resistor in place of the capacitor. The average potential at the junction of the resistor and capacitor is relatively high. In other words, V_{OUT} is almost as great as V_{IN}. The opposite occurs if the signal has a high frequency. The circuit behaves like a potential divider with a very low-value resistor in place of the capacitor. The average potential at the junction of the resistor and capacitor is therefore very low. At an extremely high frequency it is almost a short-circuit. As a result V_{OUT} is much less than V_{IN}. It may be very close to zero. The result of this is that low frequency signals appear strongly at the output of the circuit, but high-frequency signals are lost. We describe this as a **low-pass filter**. More exactly, frequencies below a certain level, known as the **cut-off frequency**, pass through the filter with very little reduction in amplitude. But there is progressive reduction in amplitude above the cut-off frequency. The way the amplitude varies with frequency is shown in Figure 18.7, p. 183.

Because it uses only passive components (see the start of this chapter), Figure 6.5a is more aptly described as a **passive low-pass filter**.

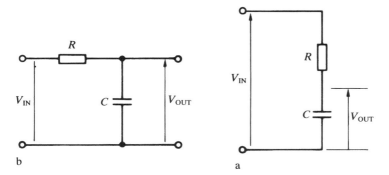

Figure 6.5 *A passive low-pass filter*

Filtering affects not only the amplitude of the signal at different frequencies but also the timing or phase of the output signal with respect to the input signal. This is because of the time factor involved in charging and discharging the capacitor. The sine wave of the output signal is still a sine wave and has the same frequency as the input signal, but its cycles begin and end a little bit later than those of the input signal. At the cut-off frequency the output signal is an eighth of a cycle behind the input signal. As frequencies increase further, the output signal lags further and further behind the input signal until it is a quarter-cycle behind.

Another type of filter is obtained if we exchange the resistor and the capacitor in Figure 6.5. Now high-frequency signals pass freely through the circuit but low-frequency signals are much reduced in amplitude. This is a **high-pass active filter**.

This simple description is only an outline of the working of resistor-capacitor filters. A filter affects not only the amplitude of the output signal but also its phase in relation to that of the input signal.

The same principles apply to a filter made from a resistor and inductor or from a capacitor and an inductor. As might be expected, a filter like that of Figure 6.5 but with an inductor instead of the capacitor has the opposite response to frequency. It is a high-pass filter.

Resonance

The circuit in Figure 6.6 is very sensitive to the effects of frequency. If the input to the circuit is a low-frequency signal, it is mainly blocked by the capacitor. But it passes easily through the inductor, which acts as a short circuit, so the signal has little effect on the circuit. Conversely, a high-frequency signal passes

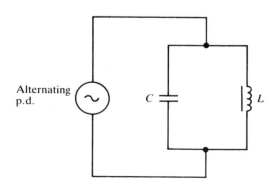

Figure 6.6   *A circuit that is resonant at a particular frequency*

readily through the capacitor but not through the inductor. Once again, there is little effect on the circuit. At one intermediate frequency, depending on the values of capacitance and inductance, the capacitor and inductor offer equal reactance to the signal. Then the signal has maximum effect on the circuit, and the swings of p.d. across it are at a maximum. When the signal source is oscillating at this frequency the circuit is said to **resonate**.

At the resonant frequency it is not necessary to supply large quantities of energy to make it oscillate. In fact, even if the signal source is removed, it will continue to resonate for a while, until the oscillations die out. The action is very like that of pushing a child on a swing. The swing has a natural frequency which depends on its length. As the child swings to and fro we give it a small push every time the child is swinging away from us. Gradually, the amplitude of swinging increases, because each time we push we are reinforcing the natural motion of the swing. When the person is at the extremes of the swinging, at the highest points above ground, energy is stored in the person as potential energy. When the person is at the lowest point of the swinging the potential energy is gone, and is converted to energy of motion. As the swing oscillates to and fro, energy is converted from potential energy to energy of motion (kinetic energy) and back again repeatedly.

A capacitor-inductor circuit is like a swing. Imagine the capacitor fully charged in a given direction. It is storing energy in the form of the electric field produced by the stored charge. Then the capacitor discharges through the inductor, losing energy itself but building up a store of energy in the inductor in the form of its magnetic field. At a certain point the capacitor is fully discharged and the inductor holds its maximum energy. Then the magnetic field begins to collapse, but self induction (p. 48) ensures that a current continues to flow in the same direction. This recharges the capacitor but in the opposite direction. Eventually the field has completely collapsed and once again all the energy is stored in the capacitor. There is repeated conversion of energy from electric field to magnetic field and back again.

If there were no loss due to resistance of the wires of the inductor and connections, and if there were no small losses in the dielectric of the capacitor, the oscillations described above would go on for ever.

Starting with a minute amount of energy in the electric or magnetic field, small oscillations will occur. If an external source supplies energy from outside at just the right frequency (like the person pushing the swing at just the right instant), a little extra energy is added to the system at each oscillation. The energy levels build up gradually until the circuit is oscillating strongly. This is the principle on which certain types of oscillators work (p. 156) and also that of the tuning circuits of radio receivers (p. 241).

7
Semiconduction

In Chapter 1 we explained the requirements for the conduction of electricity:

- There must be a supply of charge carriers, such as electrons, and ionized atoms or molecules
- The charge carriers must be free to move, electrons in the lattice of a metallic crystal, ions in solutions or in near-vacuum conditions
- There must be an electric field to provide the force (e.m.f.) to move them.

Excluding the superconductors (p. 27) below their critical temperatures, the best conductors are the metals. This is because the atomic structure of a metal is such that its lattice has a good supply of freely-mobile electrons wandering within it. Carbon is a non-metal, but this too is a good conductor. At the other extreme are the non-conductors, including most plastics and ceramics, which have no free electrons.

Between the conductors and non-conductors lies a class of material known as the semiconductors. Two of the most widely-used semiconductors are the elements silicon and germanium. Most of the descriptions which follow refer to silicon but, except where differences are pointed out, it can be taken that the description applies equally to germanium. Semiconductors are non-conductors at low temperatures but are conductors at room temperature and above. Figure 7.1 is a view of the lattice of a crystal of silicon, shown in only one plane, to make the drawing simpler. Each atom has four electrons in its outer

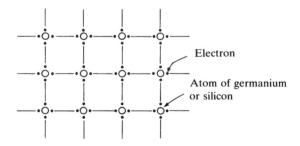

Figure 7.1 *The structure of a pure semiconductor: each atom has four electrons in its outer shell; each chemical bond (represented by a thin line) involves two electrons, one from each adjacent atom*

64

shell. The ideal number of electrons to fill this shell is eight. Each atom shares an electron with its four neighbours. Though it has only four electrons at any given instant, the shared electrons circle in the outer orbits of each atom and its neighbours, randomly going from one atom to another. At different times each atom has a full outer shell. This creates a bond between the adjacent atoms, holding the atoms together in the lattice. At room temperatures and above, a few of these electrons acquire additional (thermal) energy. Their velocity increases and they escape from their orbits. They are then free to wander in the lattice and to act as charge carriers. The semiconductor conducts, but there are few free electrons in semiconductors so they do not conduct as readily as metals.

The electrons released from the atoms of the semiconductor are known as **intrinsic charge carriers** and the conduction by these is called **intrinsic conduction**.

Resistance and temperature

Increasing temperature causes more electrons to escape from the atoms, so that the ability of a semiconductor to conduct charge increases with temperature. In other words, its resistance decreases with increasing temperature. By contrast, a conductor such as a metal already has its complement of free electrons at any temperature. Heating the metal does not release more electrons. Instead it increases the extent of the vibrations of the atoms. The vibrating atoms impede the flow of electrons through the conductor, causing its resistance to increase with increasing temperature.

Thermal runaway

A piece of semiconductor that is carrying a current becomes warmer, so decreasing its resistance. This allows the current through the semiconductor to increase, which results in further warming, further decrease in resistance, and further increase of current. This process may accelerate until the semiconductor becomes so hot that it melts. This is known as **thermal runaway**. It may be avoided by designing circuits so as to limit currents to safe levels and by providing a **heat sink**. This is a plate of heavy-gauge copper or aluminium, often with fins, that distribute the excess heat. The heat sink is bolted to or clipped on to the device, their contacting surfaces being coated with a special paste that helps the conduction of heat from the semiconductor to the heat sink.

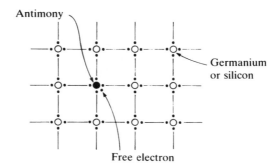

Figure 7.2   *The structure of doped n-type semiconductor: the atom of antimony has five electrons in its outer shell, so providing a spare 'free' electron*

Increased conduction

Conduction by a semiconductor is increased by doping it (see box). The effect of this is to introduce a very small percentage of dopant atoms into the lattice. Antimony is one of the elements used for doping silicon. As shown in Figure 7.2, the antimony atom has five electrons in its outer shell, one more than is found in silicon. This provides a spare electron, which is free to become a charge carrier and so to increase the conductivity of the silicon. Another element used for doping silicon is phosphorus; this too has 5 electrons in its outer orbit. Charge carriers introduced by doping are called **extrinsic charge carriers**, and conduction by these carriers is called **extrinsic conduction**. Silicon that has been doped with carriers such as antimony or phosphorus, which provide negative charge carriers (electrons), is called **n-type silicon**.

Another way of providing more charge carriers is to dope the silicon with an element such as indium or boron, which have only three electrons in their outer orbits. In Figure 7.3 we see that this results in a 'missing' electron at the site of each atom of dopant. We refer to this 'missing' electron, or 'vacancy' as a **hole**. The hole may soon be filled with an electron that has escaped from a nearby silicon atom. The escape of that electron creates a hole, which may later be filled by an electron escaping from elsewhere in the lattice. The overall result of doping is that there is a shortage of free electrons so there are more holes than free electrons.

What happens when an electric field is present is shown in Figure 7.4. There is no 'flow' of electrons. They merely jump from one atom to another atom a very short distance away. Because of the electric field, they fill a hole that is in the direction of the positive end. The effect of this is that the holes move in the opposite direction, toward the negative end. This is the direction in which positive charge carriers move, so we can consider holes to be positive charge

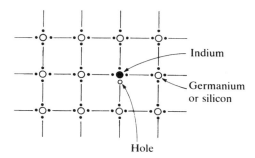

Figure 7.3 *The structure of doped p-type semiconductor: the atom of indium has only three electrons in its outer shell, so creating a vacancy or 'hole'*

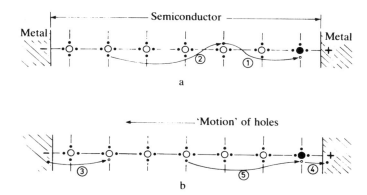

Figure 7.4 *Conduction in p-type silicon (a) At 1 an electron escapes from a silicon atom and drifts in the electric field until it finds a hole at the indium atom. At 2 the hole this created is filled by an electron escaping from an atom further along the material. The electrons drift toward the right (towards +), while the vacancies or holes pass from right to left (towards −). (b) At 3, the hole created at 2 is filled by an electron from the cell to which the material is connected, having travelled along a metal wire. At 4 an electron escapes and passes into a metal wire and on toward the cell. The hole created is again filled by an electron, so shifting the hole toward the negative end. Holes act as positive charge carriers*

carriers. For this reason, silicon doped with indium or boron is known as **p-type silicon**.

Doping

The semiconductor consists of the pure substance in crystalline form (regular arrangement of the atoms is essential) into which certain impurities (dopants) are diffused in extremely small quantities. Roughly cylindrical crystals of silicon are grown up to 30 cm diameter and 1.5 m long, then sliced into circular wafers. A whole wafer of silicon is doped by heating it in an oven and passing over it a vapour containing the dopant. The dopant condenses on to the surface of the silicon. The atoms of dopant diffuse into the silicon, eventually becoming evenly spread throughout the wafer.

The p-n junction

One of the most important effects in electronics occurs when we have n-type silicon in contact with p-type silicon. This is called a **p-n junction** . Such a junction can be made by taking a bar of n-type silicon and placing a disc of indium against half of it. Heating in a furnace causes some of the indium to diffuse into the silicon. Or, we may take a slice of n-type silicon, and heat it in a furnace with one surface exposed to boron vapour. The amount of indium or boron diffused into the n-type silicon produces enough holes to take up all the free electrons in that region of the silicon, with holes to spare. The result is that the indium-doped or boron-doped region becomes p-type silicon.

In Figure 7.5a we see a bar of silicon doped so that half is p-type and half is n-type. The bar is not connected to a circuit, so there is no electric field in it. Electrons are free to wander at random. Some of the electrons from the n-type material wander across the junction, attracted by the holes in the p-type material. The holes in the region of the junction become filled. In the p-type region (at A in Figure 7.5a) the filling of holes means that the atoms have, on average, more electrons than they should. The atoms have become negative ions. Similarly, in the n-type region the atoms have, on average, lost electrons leaving holes and so have become positive ions. Although these ions are charged,

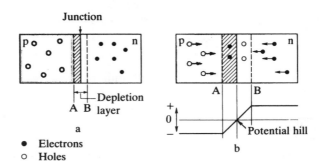

Figure 7.5   *The p-n. junction*

they can not act as charge carriers. This is because they are fixed in their places in the lattice. The result is that the region on side A of the junction has overall negative charge and the region on side B has overall positive charge. There is a p.d. across the junction. The effect of this p.d. is the same as if a cell was connected across the junction. This is not a real cell but we refer to its effects as a **virtual cell**. At a silicon p-n junction the p.d. is about 0.6 V. At a germanium junction it is about 0.2 V.

As more and more electrons wander across the junction and fill the holes on the other side, the charged regions A and B become wider. Eventually, the negative charge in region A prevents more electrons from being attracted through to the p-type material, and the charged region does not become any wider. It contains no charge carriers and, for this reason, it is called the **depletion region**.

We think of the p.d. across the depletion region as a potential hill (Figure 7.5b), the potential rising as we pass from the p-type side to the n-type side. The hill is so steep that electrons can no longer pass down it.

The situation described in the paragraph above is changed if the semiconductor is connected into a circuit and a p.d. is applied to it from an external source. We call this **biasing** the junction. In Figure 7.6a, the external source (such as a dry cell) is connected so that its p.d. is in the same direction as the p.d. of the virtual cell. It reinforces the action of the virtual cell, making the potential hill steeper. This makes the depletion region wider than before, and there is even less chance of carriers finding their way across it. No current flows across the junction, which we say is **reverse biased**.

If the external source is connected so that its p.d. opposes that of the virtual cell, the depletion region becomes narrower. The potential hill becomes less steep. The junction is **forward biased**. Electrons are more easily able to cross the junction. If the external p.d. is greater than 0.2 V (for germanium) or 0.6 V (for

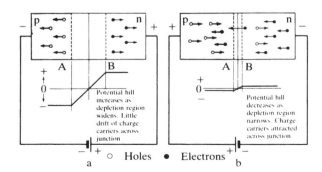

Figure 7.6 *Biasing at a p-n. junction (a) No current flows when the junction is reverse biased (b) Current flows when the forward bias is great enough to overcome the potential hill (the p.d. of the in-built virtual cell)*

silicon), the depletion region disappears and charge carriers flow freely through the material.

From this account, it is seen that the p-n junction has the unusual property that it allows conduction in one direction but not in the other.

Semiconductor diode

A semiconductor device that has a single p-n junction is known as a **diode**. The behaviour of a diode is investigated by connecting it across a variable source of p.d. and measuring the current which passes through it. Figure 7.7a shows what happens when the diode is forward biased. As the p.d. increases from zero to 0.2 V (for germanium) or to 0.6 V (for silicon) the p.d. is insufficient to overcome the virtual cell at the junction. No current flows. As the p.d. is further increased, the virtual cell is overcome and current begins to flow, as plotted in Figure 7.7a in milliamps. Current flow increases with increasing p.d., but the graph is not a straight line. This is because flow through a diode does not obey Ohm's Law (p. 20).

If the diode is reverse biased, the external p.d. reinforces the p.d. of the virtual cell and the depletion region becomes wider. Only an exceedingly small reverse current flows. This is shown in Figure 7.7b, plotted in microamps. This leakage is carried by minority carriers resulting from impurities in the semiconductors. Most diodes withstand high reverse p.d.s, of 100 V or more without any increase of leakage current. At higher reverse p.d.s, the minority carriers are strongly accelerated by the field. They gain energy and, passing close to atoms in the lattice, cause them to lose electrons. This creates a new supply of free electrons and holes. These too are accelerated strongly, interacting with the atoms to free more electrons and create more holes. The effect is

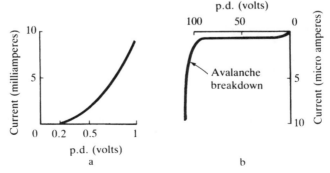

Figure 7.7 *Conduction through a diode when it is (a) forward biased (b) reverse biased. Note the difference between the current scales*

like an avalanche. The current builds up very sharply. Unless the diode is designed to withstand such a large current, it is destroyed.

The maximum p.d. that a diode can withstand when reverse-biased is known as the **peak inverse voltage** (PIV).

Diode valves

The name diode arose in the days before semiconduction was discovered, when almost all circuits were constructed using valves (vacuum tubes). One type of valve, called a diode because it contains two electrodes, has the property of one-way conduction. The name has passed to its semiconductor equivalent. Valves are rarely used today except in specialist applications such as high-power radio transmitters and certain designs of hi-fi audio amplifier.

Current through a diode

The flow of current through a forward-based diode may be summarized like this. Electrons pass from the external circuit and enter the n-type material of the diode. They flow through the n-type material as far as the p-n junction. Electrons pass from the p-type material into the external circuit, creating holes. These holes flow through the p-type material as far as the p-n junction. The holes are then filled with electrons from the n-type material. Thus although electrons enter and leave the diode, charge is carried by holes in the p-type region.

Practical diodes

Figure 7.8a shows the appearance of a typical low-power diode. It consists of a glass or plastic capsule usually a few millimetres long containing the bar of silicon (or germanium). Wires are connected to the n-type and the p-type materials, usually known as the **cathode** and **anode** respectively. The cathode end is indicated by a band marked on the capsule. These diodes are used for many purposes, wherever current must be allowed to flow in only one direction (p. 86). Typically, the current carried by such diodes is rated in milliamps or less. One special type of low-power diode is the signal diode, used in radio and TV detection circuits (p. 241) and able to respond to alternating p.d.s changing

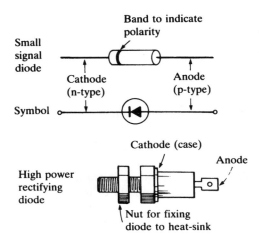

Figure 7.8   *Two typical diodes and their circuit symbol*

at radio frequencies. Signal diodes may also be used in high-speed logic circuits.

The rectifying diode in Figure 7.8 is intended to carry currents rated in amps. It has a threaded shaft at one end so that it can be bolted to a heat sink (p. 65). Such diodes are used in the rectifier sections of power-supply circuits (p. 86).

Zener diodes

Many diodes sold as 'zener diodes' are actually **avalanche diodes**. They rely on the avalanche effect described above, but are specially designed so that the avalanche action begins at a sharply defined reverse voltage, and so that the resulting high current does not destroy the diode. If we apply an increasing reverse p.d. to such a diode, there is no current (except for a minute leakage current) until the avalanche p.d. is reached, when a large current begins to flow. These diodes have applications in voltage regulator circuits (p. 88). Avalanche diodes are made so that the effect occurs at a specified reverse p.d., usually in the range 5 V to 200 V.

The true zener diodes, which have the same properties and are used in the same applications rely on the **zener effect**. The diode is made from heavily doped silicon, so that the depletion layer is very thin. Its potential hill is very steep. When the reverse p.d. exceeds a certain value (in the range 2.5 V to 5 V), electrons are able to tunnel through the depletion layer and current begins to flow.

Variable capacitance diodes

A reverse-biased diode has two conducting regions (n-type and p-type) with the depletion region between. This may be compared with the structure of a capacitor (Figure 7.9). Varying the reverse p.d. alters the width of the depletion region, and this alters the capacitance. In other words, this is a voltage-controlled capacitor. Such diodes are usually sold under the name of varicap diodes. Their capacitance is small, usually no more than a few hundred pico-farads, and they are used in certain kinds of radio and TV tuning circuits. They have the advantage that the tuning can be controlled electronically instead of by turning a tuning knob.

Light-emitting diodes and laser diodes are described in Chapter 13

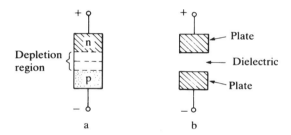

Figure 7.9 *Comparing (a) a reverse-biased diode with (b) a capacitor.*

8

Transistors

Transistors of various types form the basis of most electronic circuits. As we shall explain later, there are two fundamental applications for transistors – as electronically controllable switches and as amplifiers. We describe several different types of transistor in this chapter, beginning with the field effect transistor, which is the easiest to understand.

Field effect transistor

Strictly, this should be described as a **junction field effect transistor**, often shortened to JFET. This name comes from the fact that its action depends on what happens at a p-n junction (p. 68). The structure of a JFET is shown in Figure 8.1a. A bar of n-type semiconductor has metal contacts at each end. On either side of the bar is a layer of p-type material, the two layers being con-

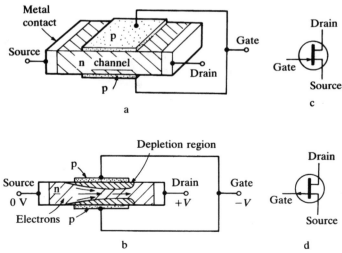

Figure 8.1 *Junction field effect transistors (JFETs) (a) Structure (b) How the increasing depletion region makes the channel narrower (c) Symbol for an n-channel JFET (d) Symbol for a p-channel JFET*

nected together by a fine wire. Remember that this structure measures only a millimetre or so across.

If a p.d. is applied to the ends of the bar, a current flows along the bar. Because the bar is made from n-type material, the current is carried by electrons. We say that this is an **n-channel JFET**. In referring to the potentials in this diagram we take the potential of the source to be zero (see box).

Electrons enter the bar at the **source** terminal (the reason for its name), and leave at the positive end of the bar, the **drain** terminal. The layers of p-type material are known as the **gate**. If these are at 0 V or a slightly more positive potential, they have no effect on the flow of electrons. But if the gate is made more negative than the source (perhaps by connecting a cell with its positive terminal to the source and its negative terminal to the gate), an important effect follows. The p-type gate and the n-type bar form a p-n junction. With the gate negative of the bar, the junction is reverse biased and a depletion region (p. 68) is formed (Figure 8.1b). Putting it another way, the negative charge on the gate repels electrons as they flow along the bar. The channel through which they flow is made narrower. In effect, the resistance of the bar is increased, and this reduces the flow of electrons through the bar. If the p.d. between the gate and source is very high, the depletion region extends across the whole width of the bar. This cuts off the flow of electrons altogether. Used in this way the JFET is a voltage-controlled switch.

Zero potential

We often mark the terminals of a cell or battery with '−' and '+' to indicate the negative and positive terminals. The negative terminal is the one from which negative charge carriers leave the cell. The other terminal is said to be more positive than this, and it is to this terminal that the electrons flow. When we refer to potential the terms 'positive' and 'negative' have slightly different meanings. They describe the potential of a point in space, or, in more practical terms, of a terminal or an electrode, or any other point in a circuit. In these terms we can say that one point is 'more positive than' or 'more negative than' some other point. The terms 'positive' or 'negative' indicate whether the potential goes up or down as we go from one point to another. Potential can only be described relative to the potential at some other point.

From the definition of potential (p. 17), zero potential would be the potential of some point in space far away from any charged bodies. Even if such a point exists it would not be practicable to base our measurements on this. We need a reference point which is closer to hand. Often we take the potential of the Earth as our reference point. We may refer to this as **ground potential** or simply

ground, and rate this as being 0 V. All potential in circuits connected to ground (that is, earthed circuits) are measured with respect to this.

In circuits that are not actually earthed, it is usually most convenient to take one of the conductors that runs to all or most parts of the circuit as the reference point. Often this conductor is connected to the negative terminal of the cell or other power supply. Although it may not be earthed, this conductor may still be referred to as 'ground'. Other points may have potentials positive to this; others may be negative, as in Figure 8.1.

With smaller p.d.s that do not completely cut off the current, the width of the bar left free for the passage of electrons is proportional to the p.d. As p.d. increases, so the width decreases, the resistance increases and the current decreases. In effect the JFET is a voltage-controlled resistor.

A JFET is always operated with the p-n junction reverse-biased, so current never flows from the gate into the bar. The current flowing into or out of the gate is needed only to change the potential of the gate. Since the gate is extremely small in volume, only a minute current (a few picoamps) is required. A small change in potential of the gate, and hence an extremely small current, controls the much larger current flowing through the channel. The transistor is acting as an amplifier.

JFETs have many applications, particularly in the amplification of potentials produced by devices such as microphones which are capable of producing only very small currents. They are also useful in potential-measuring circuits such as are found in digital test-meters (p. 161), since they draw virtually no current and therefore do not affect the potentials that they are measuring (p. 162).

A JFET similar to that described above is manufactured from a bar of p-type material with n-type gate layers. Current is conducted along the bar by holes and this is known as a p-channel JFET. In operation, the gate is made positive of the source to reverse-bias the p-n junction.

MOSFETs

The action of a field effect transistor depends upon the effect of the field produced in the region of the charged gate. Another type of FET is illustrated in Figure 8.2a. This is the MOSFET which is the acronym for **metal oxide silicon field effect transistor**. The transistor is based on a bar of p-type silicon. Two crosswise strips are doped to make them into n-type material, and metal is deposited on these to form the source and drain terminals. Between the strips,

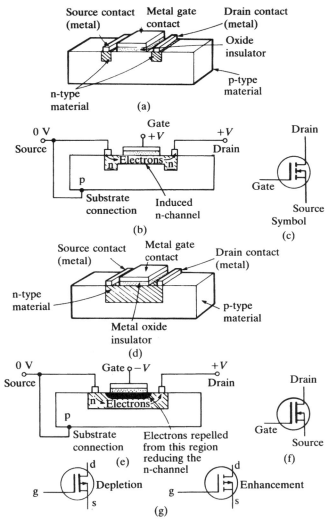

Figure 8.2 *Metal oxide silicon field effect transistors (MOSFETs) (a–c) n-channel enhancement MOSFET; (d–f) n-channel depletion MOSFET (g) Symbols for p-channel MOSFETs*

the surface of the silicon is covered with a thin layer of silicon oxide, which is a non-conductor. Metal is deposited on the silicon oxide to form the gate electrode.

The action of this transistor is similar to that of the JFET, and is illustrated in Figure 8.2b. A potential is applied between the source (0 V) and the drain (positive) but there is a p-n junction between them and the p-type material, so no current can flow. It is like having two diodes connected back-to-back. The p-type material is connected to the source so it is at 0 V. When the gate is made positive of the p-type material, it repels the holes from nearby regions of the p-type material, turning it temporarily into n-type. Another way of looking at this is to say that the gate attracts electrons to the region around it. These are more than enough to fill the holes in that region and the remainder are available as charge carriers. This creates a channel joining the two n-type strips. Current can then flow from the source to the drain. The greater the gate potential, the wider the channel and the larger the current.

This type of transistor is made to conduct by applying a positive voltage to the gate. Compare this with the JFET which requires a negative voltage on the gate. As in the JFET, no current flows from the gate into the other parts of the transistor and only a minute current is required to vary the potential of the gate. This kind of MOSFET is probably the most widely used but there are other sorts.

The transistor in Figures 8.2a and b is known as an n-channel enhancement MOSFET. The term 'enhancement' refers to the fact that there is no channel when the gate is at 0 V, but the channel is enhanced (made wider) as the gate potential increases. The transistor of Figures 8.2d and e is an n-channel depletion type. This has a channel of n-type material connecting source to drain so current flows even when gate is at 0 V. If the gate potential is reduced below zero, electrons are repelled from the region close to the gate. The channel is depleted of electrons, so reducing the width of the channel and restricting current flow (Figure 8.2d). If the gate potential is increased, electrons are attracted from the p-type material, making the channel wider and increasing current flow. So the conduction is affected by both negative and positive potentials of the gate.

Gate potential

If the gate of a MOSFET is left unconnected, electric fields from outside can easily charge it. The small charge acquired in this way is enough to produce a significant effect on conduction through the transistor. When a person is constructing a circuit there are many ways in which fields can be generated and the gate may be affected even without coming into direct contact with charged surfaces. This makes the behaviour of MOSFETs erratic when the gate is unconnected. No gates should be left unconnected.

A problem may arise when MOSFETS (and JFETS) are handled during circuit construction. The human body becomes charged due to friction between the body and clothing, or floor coverings. Usually we do not notice this because although the potential of the body may rise to several hundreds or thousands of volts, the amount of charge is small and the currents produced when we touch grounded objects are usually (though not always) too small to be felt. But if we touch a gate terminal of a MOSFET, the discharge current can easily penetrate the very thin layer of oxide and destroy the transistor. Precautions must be taken to avoid this happening. These include storing MOSFETS in packages made from conductive foil, or with their terminal wires pushed into conductive plastic foam, not wearing clothing made from synthetic fibres, and wearing an earthed wrist-band.

In addition to the n-channel MOSFETS there is a range of p-channel MOSFETS, including both enhancement and depletion types. Their structure and action is equivalent to that of n-channel MOSFETS, except that the polarities are reversed. Integrated circuits containing both n-channel and p-channel enhancement mode MOSFETS are the basis of CMOS logic (p. 190).

VMOS

These are sometimes known as v-channel MOSFETS, but here we are referring to the shape of the channel, not its charge-carriers. JFETS and MOSFETS of the structure shown in Figures 8.1 and 8.2 are unable to carry heavy currents. In VMOS transistors (Figure 8.3) the doped layers are built up on a surface that has previously been etched to form steep-sided grooves, v-shaped in section. The action of the n-channel enhancement-type transistor in Figure 8.3 is the same as in Figure 8.2b. The difference is that conduction is down the sides of the grooves. The distance from top to bottom of the channel is short, but the

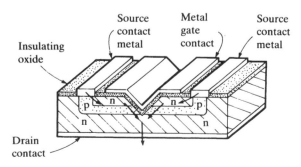

Figure 8.3 *A VMOS transistor: arrows show the flow of electrons when the n-channel is generated*

channel is very wide in the direction along the grooves, so its resistance is low. This means that currents can be large without overheating the transistor. VMOS power transistors are made to carry currents of 10 A or more. They make it possible to control large currents by devices that can produce only a small current. For example, a heavy-duty relay (p. 118) or a 10 Ω loudspeaker can be operated directly from the output of a CMOS logic gate.

HEXFETS

Another type of structure is the HEXFET, which is an n-channel enhancement device. It is based on a hexagonal pattern repeated nearly 80 000 times in a square centimetre (Figure 8.4). The gate is a hexagonal network made of silicon, surrounded by an insulating layer of silicon oxide. The hexagonal pattern of the gate is matched by the pattern of the diffused n-type layer. When a positive charge is applied to the gate, the p-type layer around the border of each hexagonal cell is converted to n-type and a current flows. Conduction is mainly upward through the body of the transistor, a short distance over a large area, so resistance is low. The n-channel offers little resistance as there are tens of thousands of hexagons in a relatively small chip area and the width of each border area is small. As a result of this structure the 'on' resistance of the transistor is around one third of that of MOSFETs of equivalent size. This

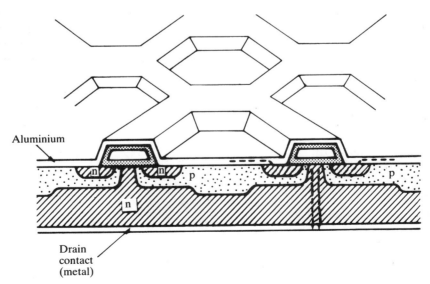

Figure 8.4   *A HEXFET: the arrows show the direction of flow of electrons when the n-channel has been generated*

makes it possible to switch large currents with relatively little loss of voltage across the transistor and minimal dissipation of energy due to transistor resistance, reducing the risk of overheating.

One special type of MOSFET is the **current-sense power MOSFET**. In order to allow a large current to flow though the transistor it is configured so that it is, in effect a very large number of small transistors connected in parallel (see HEXFET, p. 80). The sources of these transistors consists of a single layer of aluminium. In a current-sense MOSFET, a small section of this layer is isolated from the remainder. This creates a few transistors in parallel with each other and with the majority. There is a connection from this separate source area to a special current-sense pin. Together, the majority conduct the very large current through the power transistor, from the source to the drain. The few isolated transistors conduct a much smaller current between their separate source (the current-sense pin) and the common drain of the transistor. The current flowing into the current-sense pin is very much smaller than that flowing into the source pin, but the two currents are always proportionate. We use external circuitry to measure the current flowing into the current-sense pin and use this measurement to find the current flowing into the source pin. In this way we measure the current passing through the power transistor without affecting the circuit in which it is flowing.

Bipolar transistors

Owing to practical difficulties in making FETs in the early days of semiconductors, bipolar transistors were the first to be widely used. They still are extremely popular with designers, even though the problems of making FETs have been overcome. They work on an entirely different principle to FETs.

A bipolar transistor is a three-layer device consisting either of a layer of p-type sandwiched between two n-type layers or a layer of n-type between two p-type layers. These are referred to as n-p-n and p-n-p transistors respectively. The fact that conduction occurs through all three layers, which means that it involves both electrons (negative) and holes (positive) as charge carriers, is why these are called **bipolar transistors**. Figure 8.5 shows the theoretical structure of an n-p-n transistor, the type which is most commonly used. The diagram shows the 'sandwich' structure. This is less obvious in Figure 8.6, which shows the way the n-p-n transistor is built up on a silicon chip (see box).

Mass production

Very large numbers of transistors are produced at once on a single wafer of silicon (p. 66). A popular technique for this is the planar process. First a layer of silicon oxide is formed on the upper surface of the wafer by heating it in an atmosphere of oxygen and water vapour. The next step is similar to that used for making pcbs (p. 134), except that it is on a very much smaller scale. The silicon oxide layer is coated with a layer of photoresist. Then the wafer is exposed to uv light, usually through a mask on which the areas to be etched are left clear. The pattern of the mask is repeated over and over again to produce many transistors, which can be later separated by cutting the wafer. Another approach is to have a single mask and to step this along to produce an array of exposed areas. For finely detailed circuits, a reduced image of the mask is projected onto the photoresist layer using a lens system.

The resist is etched chemically, removing the exposed areas of the resist and the oxide beneath. Finally, the wafer is doped, as above, but with a different dopant. This time the dopant can not reach the areas covered with oxide, so only the exposed areas are doped. It is possible to control the depth to which the dopant penetrates. When a transistor such as that in Figure 8.6 is made, we begin with a wafer doped to produce n-type silicon. Areas of this are etched and doped to produce the p-type silicon of the base layer. Although the silicon in the base begins as n-type, it is converted to p-type by diffusing an excess of a hole-producing dopant, to cancel out the effects of the electron-producing dopant already present. The wafer is then exposed to a mask which causes smaller areas to be etched over the base layer, and then the wafer is doped to produce a smaller shallower n-type layer (emitter) in the base layer. Finally the collector (substrate), base and emitter contacts are added by placing the wafer in a vacuum, covering it with a mask and evaporating metal on to it. Alternatively a continuous layer of metal may be deposited and etched away to leave the required connections.

The three layers of the transistor are known as the **collector**, the **base** and the **emitter**. In effect, the transistor consists of two p-n junctions, or diodes, connected back-to-back. It would seem that it is impossible for current to flow from the collector to the emitter or from the emitter to the collector for, whatever the direction of the p.d., one or other of the p-n junctions is sure to be reverse-biased (p. 69). This is where the base layer is involved. One point about this that is not shown in Figure 8.5 is that it is very thin. Also, it is lightly doped, so it provides very few holes.

When the transistor is connected as in Figure 8.5, the base-emitter junction is forward-biased. Provided that the base-emitter p.d. is greater than about 0.6 V (p. 70), a base current flows from base to emitter. But, when putting it this way, we are describing it in terms of conventional current (p. 12). What actually

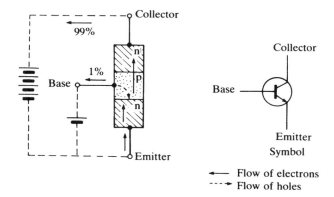

Figure 8.5   *The action of a bipolar junction transistor (BJT)*

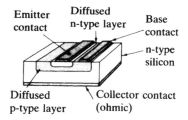

Figure 8.6   *The structure of a diffused n-p-n junction transistor*

happens is that electrons enter the emitter and flow to the base-emitter junction. Then, as in a diode (p. 69), they combine with holes that have entered the base terminal. As there are few holes in the base region, there are few holes to fill. Typically there is only one hole for every 100 electrons arriving at the base-emitter junction. The remaining 99 electrons, having been accelerated toward the junction by the field between the emitter and base are able to pass straight through the thin base layer and also through the depletion layer at the base-collector junction, which is reverse-biased. The electrons flow on toward the collector terminal, attracted by the much stronger field between emitter and collector. They flow from the collector terminal, forming the collector current, and on toward the battery. In effect, the base-emitter p.d. starts the electrons off on their journey but, once they get to the base-emitter junction, most of them come under the influence of the emitter-collector p.d. The collector current is about 100 times greater than the base current. We say that there is a **current gain** of 100.

If the base-emitter p.d. is less than 0.6 V, the base-emitter junction is reverse-biased too. The depletion region prevents electrons from reaching the junction.

The action described above does not take place and there is no collector current. In this sense the transistor acts as a switch whereby a large (collector) current can be turned on or off by a much smaller (base) current. If a varying current is supplied to the base, a varying number of electrons arrive at the base-emitter junction. The size of the collector current varies in proportion. In this sense the transistor acts as a current amplifier. The size of a large current is controlled by the variations in the size of a much smaller current. These functions, switching and current amplifying are the two chief functions of transistors.

The structure and operation of p-n-p transistors is similar to that of n-p-n transistors, but with polarities reversed.

Practical transistors

A transistor is a minute object on a small chip of silicon (rarely germanium). To make it practicable to handle the transistor it is mounted in a case or sealed into a block of plastic, with thin wires connecting the base, emitter and collector to thicker terminal wires. Figure 8.7 shows typical transistor packages. These are standard packages also used for JFET and MOSFET transistors.

By varying the amount of doping, the method of doping, and the geometry of the regions, transistors with various characteristics can be made. Some have much higher gain than others (up to 800 times), or may be suitable for operation at higher p.d.s. In typical general-purpose transistors the maximum collector current is only a few hundred milliamps. By contrast, power transistors are capable of collector currents of 30 A. Other types are designed specially for

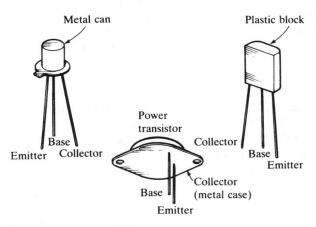

Figure 8.7   *Types of transistor package*

operation at high frequencies. At radio frequencies the capacitance between the base and emitter may act to reduce the amplitude of the signal. This is known as the **Miller effect**. Radio-frequency transistors are designed to minimize this effect.

Darlington transistors

A Darlington pair consists of two n-p-n transistors connected as shown in Figure 8.8. The way they work is easier to understand if it is explained in terms of conventional current. Given a small base current flowing to TR_1, a much larger collector current flows into the transistor and out of the emitter. This becomes the base current of TR_2, which amplifies the current still further. For small base currents, the gain of the Darlington pair equals the gain of TR_1 multiplied by the gain of TR_2. A typical Darlington pair has a gain of 10 000, and some have gains up to 50 000. A Darlington pair may be assembled from two individual transistors but it is more convenient to use a **Darlington transistor**. This consists of two n-p-n transistors made as one unit, with the connections shown in Figure 8.8. (not the resistor) and packaged in one of the standard transistor packages shown in Figure 8.7. Paired FETs are available, often under the name **FETlington**.

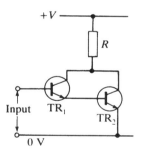

Figure 8.8 *Two n-p-n transistors connected as a Darlington pair*

9
Semiconductor circuits

This chapter describes ways in which diodes, transistors and some of the other semiconductor devices are used.

Rectifiers

Almost all electronic circuits need a d.c. power supply. The action of an n-p-n transistor requires the collector to be positive of the emitter, and does not occur if the polarity is reversed. An alternating power supply is of no use. The same applies to other transistors and related devices. Batteries provide a d.c. supply and are particularly suitable for portable equipment, but, for fixed equipment, it is more economical in the long run to use the alternating supply provided by the mains. Normally this has far too high a voltage (230 V) for use with electronic circuits. This is no problem because, since the current is alternating, it can be stepped down to a suitable level with a transformer (p. 55). The output of the transformer is a low-voltage alternating supply. The next requirement is to convert the a.c. supply into a d.c. supply. This process is known as **rectification** and a circuit which does this is called a **rectifier**.

Single-diode rectifier

The obvious choice for a rectifying device is a diode, for this allows current to flow through it in only one direction. Any diode can be used, but special rectifier diodes are made, suitable for withstanding the currents and reverse p.d.s in rectifier circuits. Figure 9.1 shows the simplest kind of rectifier circuit. On the left is a step-down transformer rated to produce the required voltage from the mains supply. On the right is a single diode which performs the rectification. To understand what happens, look at the graph of Figure 9.2a. The waveform of the a.c. from the transformer has the shape of a sine wave. If it is taken from the mains supply, it has a frequency of 50 Hz. One complete wave takes a fiftieth of a second so the total time for the three cycles shown in Figure 9.2a is three-fiftieths of a second or 0.06 s.

During the first half of each cycle the waveform is positive. This means that current flows out of terminal A (Figure 9.1) of the transformer, through the

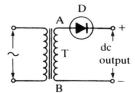

Figure 9.1 *A single-diode rectifier*

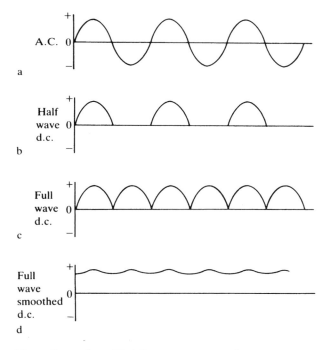

Figure 9.2 *Alternating and rectified direct current waveforms*

diode, through any circuit that happens to be connected, and back into the transformer coil through terminal B. During the second half of the cycle, current would flow out of B and into A but this is prevented because the diode will not allow it to flow in this direction. The result is shown in the graph of Figure 9.2b. The diode conducts during the first half of each cycle and a pulse of current flows around the circuit. No current flows during the second half of the cycle.

This type of rectifier produces one-way current but has the disadvantage that it supplies current for only half the time, so it is inefficient. It is known as a **half-wave rectifier.**

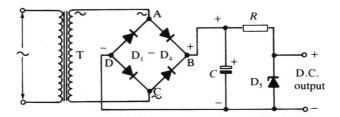

Figure 9.3 *A bridge rectifier, with a smoothing capacitor and with a zener diode for regulating its output voltage*

Bridge rectifier

Figure 9.3 shows a circuit that produces current during both halves of the cycle. It consists of four diodes connected in a bridge (compare with p. 60). During the first half of the cycle, current flows from the transformer to A, through the diode to B, round the circuit from B and back to D, through the diode to C and finally back to the transformer. The diodes between A and B and between D and C are conducting, but the other two diodes are reverse-biased so do not conduct. During the second half of the cycle current flows from the transformer to C, through the diode to B, round the circuit from B and back to D, through the diode to A and finally back to the transformer. The diodes between C and B and between D and A are conducting, but the other two diodes are reverse-biased so do not conduct. This produces the waveform shown in the graph of Figure 9.2c, and the diode bridge is known as a **full-wave rectifier**. A bridge rectifier may be assembled by connecting four individual diodes, or a rectifier bridge unit consisting of four diodes sealed into a single four-terminal package may be purchased.

Smoothing

The output of half-wave and full-wave rectifiers is pulsed d.c., quite unlike that obtained from a battery. It is suitable for powering some kinds of circuit, but not for all. If a pulsed supply is used with an audio circuit, for example, the result is a strong 'mains hum' at 100 Hz (double the mains frequency) super-imposed on the audio signal. Similarly, a pulsed supply is unsuitable for power-ing logic circuits. The solution is to connect a large-capacitance capacitor across the output from the rectifier. Capacitances of several thousand micro-farads are required so we normally use an electrolytic capacitor, as shown in Figure 9.3. During the peak of each half-cycle the capacitor receives a boost of charge from the rectifier. The charge passes out to the powered circuit at a steady rate so there is a slight fall of voltage after each peak. If the capacitance is sufficiently large, voltage will have fallen by only a small amount before the capacitor is recharged by the next peak. As shown in the graph of Figure 9.2d, the waveform of smoothed d.c. is a more-or-less steady voltage with a slight ripple at 100 Hz.

Voltage regulation

Mains power supply units (often made so as to plug into an ordinary mains socket) are sometimes sold as 'battery eliminators' . They are used to provide power for running a radio set or tape recorder, and are more economical than using batteries. Their output is d.c. and may be rated at 6 V, 9 V and sometimes other set voltages. If you measure the output using a test-meter, you may find that the voltage is much higher than the rated voltage when nothing is connected to the unit. When a radio set or other piece of equipment is connected the output falls to the rated level, but may possibly be several volts lower. This is because the voltage of a simple transformer-rectifier-smoother falls as the current drawn from it increases. This does not usually matter for radio sets and similar equipment but it does matter if parts of a circuit are designed to operate at a fixed voltage.

If it is necessary to regulate the voltage from a power supply, an additional stage is required. In Figure 9.3, the additional stage is made up from a resistor and a zener diode. The zener diode is chosen to have a zener voltage (p. 72) equal to that required by the circuit which is to be supplied. This must be lower than that provided by the rectifier circuit. When current is flowing to the external circuit there is a p.d. across the resistor. The value of this is chosen so as to bring the voltage down to slightly more than that required by the external circuit when it is using its maximum current. The zener conducts a small surplus current and the remainder goes to the external circuit. The p.d. across the d.c. output terminals is the zener voltage. If the external circuit changes its requirements so that it needs less current, the excess current flows away through the zener diode. The supply to the external circuit remains at the zener voltage. Regulation by a zener diode is not perfect but is adequate for many purposes.

A **bandgap voltage reference** has a similar action to that of the zener diode. On p. 105 it is explained how a bandgap device may be used as a temperature sensor, by adjusting the rates of change of two opposing p.d.s. In the bandgap voltage reference the adjustments are such that the same voltage difference is obtained at all temperatures within a wide range. The reference therefore gives a constant voltage at any temperature, making it suitable for precision circuits. It can replace the zener diode in Figure 9.3, to produce a better degree of regulation of the output voltage.

Power control

Figure 9.4 shows a half-wave rectifier employing a thyristor (p. 100). This does not conduct at all unless a positive trigger pulse is applied to its gate. When triggered, and assuming that the output of the transformer is in the first half-cycle, it behaves like the diode of Figure 9.1. But as the current dies away beyond the peak, it eventually falls below the holding current and conduction stops. It does not start again unless the thyristor is triggered during the first half of the next cycle.

In thyristor circuits the pulse is applied automatically once during each positive (first) half-cycle. The pulse generating circuit allows the pulse to be generated during any desired stage of the half-cycle, Figure 9.5 shows what happens. Here the pulse is applied just as the a.c. voltage reaches its peak. The thyristor begins to conduct and the output is as shown on the bottom graph. As the a.c. voltage falls to zero and goes negative the thyristor becomes non-conducting. It does not conduct again until it is triggered half-way through the next half-cycle. As a result, current flows for only a quarter of a cycle. By altering the timing of the trigger pulse we can arrange for conduction to com-

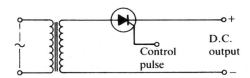

Figure 9.4   *Using a thyristor (silicon controlled rectifier)*

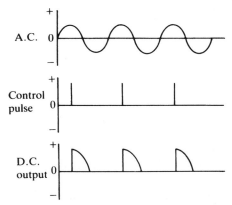

Figure 9.5   *Alternating current and the controlled direct current as produced by a thyristor circuit*

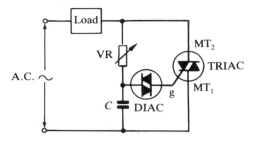

Figure 9.6 *A full-wave power control circuit using a thyristor and diac*

mence earlier or later in the cycle. In this way the average current flowing to the external circuit can vary between the maximum (equivalent to half-wave d.c.) and zero. Circuits of this type are in common use for controlling the brightness of room lights and the speed of motors on equipment such as electric drills. In the case of equipment operating at mains voltage there is no need for the transformer.

The thyristor shares the disadvantage of the single diode, that it can only conduct half the time. The **triac** (p. 102) is the solution to this problem. It can be triggered in each half of the cycle, so the power obtainable can be varied between zero and that of full-wave d.c. A simple triggering circuit is shown in Figure 9.6. The alternating potentials in the circuit result in rapidly changing potentials between one plate of capacitor *C* and the variable resistor VR. By altering the setting of VR, the timing of the charging and discharging of *C* may be adjusted. This allows trigger pulses to be generated at any required stage during the a.c. cycle. Triggering is effected by a **diac** (p. 102). When the p.d. across it exceeds its breakdown voltage, the avalanche effect produces a sharp positive pulse which triggers the triac.

Transistor connections

There are three basic ways in which an n-p-n bipolar junction transistor may be connected and used in a circuit (Figure 9.7). In the left-hand drawing the emitter (e) is connected to both the input and the output sides of the circuit. It is common to both sides, so this is known as the **common emitter connection**. Similar reasoning gives the names to the **common base** (b) and **common collector** (c) connections. The common emitter connection is the most frequently used, so we look at it first.

Figure 9.8 shows a transistor in the common emitter connection. When the switch S is closed, a small base current flows to the transistor and a larger current flows in through the collector and out through the emitter. There is a

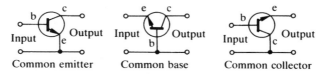

Figure 9.7   *The three basic ways of connecting a transistor*

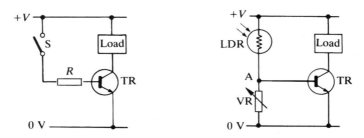

Figure 9.8   *Using an n-p-n transistor as a switch, in the common-emitter connection*

Figure 9.9   *A variation of the switching circuit of Figure 9.8; the circuit is sensitive to light falling on the LDR and switches a load on or off accordingly*

load between the collector and the positive supply line, so the large current flows through this. The load might be a lamp, or a motor or the coil of a relay. Closing S lights the lamp, starts the motor or energizes the relay. Normally there would be no point in simply having a switch to turn on the base current; the circuit of Figure 9.8 is more suitable as a demonstration of transistor switching action. A more useful variation of the circuit appears in Figure 9.9. Two resistors, a light-dependent resistor LDR (see p. 120) and a variable resistor VR form a potential divider (p. 57). The potential at point A depends on the setting of VR and on the amount of light falling on the LDR. We can set VR so that the potential at A is just not enough to produce a base current to turn the transistor on. Then, if the amount of light increases, the resistance of LDR decreases and therefore the potential at A rises. If it rises sufficiently enough base current flows to saturate the transistor and turn it on. If the transistor is turned on, the load (for example, a motor) is also turned on. The circuit turns the motor on by daylight but turns it off in darkness. The precise light level at which the motor is turned on can be set by adjusting VR.

Load p.d.

When the transistor in Figures 9.8 and 9.9 is off, no current flows, so there is no p.d. across the load (see box, p. 24). This means that the potential at the collector is the same as the supply voltage ($+V$). When the transistor is on, current flows through the load and causes a p.d. to develop across it. This means that the potential at the collector falls. The bigger the collector current the more the collector potential falls. The fall reaches its maximum when the transistor is saturated, and the collector potential is about 0.7 V.

Protective diode

A precaution must be taken when using a transistor to switch an inductive load, such as a relay, a solenoid or a motor. When the transistor switches off, the current through the load ceases instantly and the magnetic field in the coil collapses. The effects of self-induction cause a very large e.m.f. to be generated (p. 52), so as to attempt to make the current continue to flow. This causes a p.d. to be produced across the load which is very much higher than the p.d. originally present. It could easily damage the transistor. To prevent this we connect a diode across the load (Figure 9.10). Current does not normally flow through the diode, as it is reverse-biased. But, when self-induction causes the large current to flow down through the load from the positive rail, the diode diverts this current back to the positive rail, and it does not flow into the transistor.

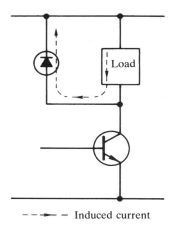

– – ▸ – – Induced current

Figure 9.10 *Using a protective diode with an inductive load*

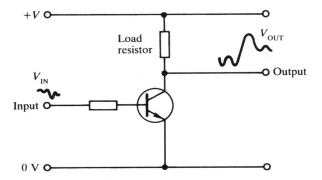

Figure 9.11   *A common-emitter amplifier. The input signal is amplified and inverted.*

The example above shows the transistor being used as a switch. We could exchange the LDR and VR to make a circuit that turns the load off in bright light and on in dim light. Many other types of sensor could be substituted for the LDR. The other use for a transistor in the common emitter connection is as an amplifier (Figure 9.11). In this circuit, the load is a resistor. The behaviour of this circuit is similar to that of the switch except that its important action occurs in the region between turning on and the onset of saturation. As the base potential increases from 0.6 V and increased base current flow, an increased (and larger) collector current flows, an increasing p.d. develops across the load and the collector potential falls further below the supply voltage. The potential at the collector is taken to be the output of the circuit. Just as a small base current leads to a large collector current, so a small change in input (base) voltage produces a relatively large change in output (collector) voltage. The circuit acts as a voltage amplifier, but collector potential goes down as base potential goes up, so it is an **inverting amplifier**.

Common base connection

A typical use of the common base connection is to build a voltage regulator for use as the regulating stage of one of the power supplies described earlier in this chapter. From the unregulated side of the circuit (Figure 9.12), current flows through a resistor and a zener diode. The zener diode holds the base of the transistor at its zener voltage (V_Z) by its regulating action (p. 89). The transistor is usually rated to pass currents of 1 A, possibly more. Current flows through the collector and emitter to the external circuit. As long as the transistor is conducting, there is the usual p.d. (V_{BE}) of about 0.6 V between the base and the emitter. So the regulated output voltage V_{REG} equals the sum of

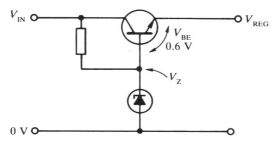

Figure 9.12 *A transistor in the common-base connection being used as a voltage regulator.*

V_Z and V_{BE}. For example, if the zener voltage is 4.7 V, the regulated output voltage is 5.3 V.

V_{REG} is relatively unaffected by changes in the current being drawn by the circuit that it is supplying. If the current increases, this tends to increase the base-emitter p.d. (since V_Z is held constant) but this instantly causes the collector current to increase, supplying extra current to the circuit until the base emitter p.d. settles to 0.6 V again. The reverse action occurs if the amount of current being drawn is decreased. V_{REG} is similarly unaffected by changes in V_{IN}.

Voltage regulators

A transistor circuit such as the above may be incorporated into a type of integrated circuit (p. 136) known as a **voltage regulator**. Such devices may also have current limiting features. If the load develops a short circuit or if the current drawn exceeds a safe amount for any other reason, this condition is detected by the regulator and the voltage output is sharply reduced. The circuit may also include a thermistor, which provides thermal shutdown, cutting off the current if the device becomes overheated. Voltage regulators are made to provide one fixed voltage in a standard range, including 5 V, 6 V, and 12 V. Variable regulators are also available.

Common collector connection

Figure 9.13 shows a frequently used application of the common collector connection. The difference between V_{IN} and V_{OUT} is the typical base-emitter voltage of a silicon transistor, 0.6 V. Whatever the value taken by V_{IN}, the value of V_{OUT} is always 0.6 V less. The emitter potential follows the input, so it is known as an **emitter follower** amplifier. This means that, as an amplifier, this

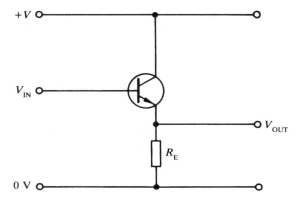

Figure 9.13   *The common-emitter connection being used as an emitter-follower amplifier*

circuit has a voltage gain of one. There may not seem much point in having an amplifier with a gain of one, but the circuit has its uses. The base current is, like all base currents, very small. The circuit takes very little current to drive it. It can be driven from another circuit which is able to supply only a very small current. An example is a crystal microphone (p. 107) which produces current of only a few microamps. But the common emitter amplifier is able to supply a very much larger current, its collector current, to anything connected to V_{OUT}. It has a high **current gain**, often greater than 100.

Note that, as input potential goes up, so does output potential. This is a **non-inverting amplifier**.

Schmitt trigger

One of the drawbacks of the common emitter transistor switch (Figure 9.9) is that it switches on and off gradually. If the light level changes slowly, as it often does, the switch spends perhaps ten minutes or more in an intermediate state. Enough base current is flowing to produce a collector current but not enough to fully saturate the transistor. So the load is switched half on and may waver in its action, especially if the light level is fluctuating slightly. We need a switching circuit that has a more definite 'snap' action. This is provided by the Schmitt trigger circuit (Figure 9.14).

This description of the way this circuit works begins with the input at a potential less than 0.6 V. TR_1 is off so its collector potential at A is almost equal to $+V$ (see above). A small current flows through R_1 and R_2 to the base of TR_2, turning on. The large collector current of TR_2 flows through the LED (illuminating it) and also through R_3. Since a current is flowing through

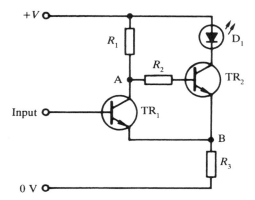

Figure 9.14 *A Schmitt trigger circuit*

R_3, there is a small positive potential at point B. For the sake of the discussion, assume that this is 1 V.

If the input potential is gradually increased, nothing happens until the base-emitter p.d. of TR_1 exceeds 0.6 V and a base current begins to flow. But, since point B (and therefore the emitter of TR_1) is already at 1 V, the base of TR_1 must be raised to 1.6 V before TR_1 begins to turn on. When TR_1 begins to turn on, the potential at A begins to fall. This reduces the base current to TR_2. Its collector current is reduced. The current through R_3 is reduced and so is the p.d. across R_3. The potential at B begins to fall.

As the potential at B falls, so the base-emitter p.d. of TR_1 begins to increase. This is because the base is at 1.6 V but the potential at B is now less than 1 V. This increasing base-emitter p.d. turns TR_1 on even more strongly. Its collector current increases further, decreasing the potential at A and turning TR_2 further off. Once begun, the action continues more and more rapidly until TR_1 is fully saturated and TR_2 is fully off. The LED goes out. It can be seen that only a very small rise of input above 1.6 V is enough to trigger off the events described above.

However, a correspondingly small fall of input below 1.6 V does not have the reverse effect. Now that TR_2 is off, there is no current through R_3 and point B is at zero. To make the base-emitter p.d. of TR_1 less than 0.6 V, the input must be brought below 0.6 V. The input has to fall by 1 V or possibly more. When this happens, the reverse action occurs, TR_1 turns off and TR_2 and the LED turn on again.

To summarize the action, the circuit is triggered when the input increases slightly above the **upper threshold** (1.6 V in the example). Once triggered, it is not triggered in the reverse direction until the input has fallen below the **lower threshold** (0.6 V in this example). One feature of the action is that only a slight increase or decrease at the appropriate threshold is enough to initiate an abrupt change in the state of the circuit. The other feature is that the thresholds are relatively far apart. The exact values of the thresholds and their distance apart may be set by choosing suitable values for the resistors. The circuit is shown operating an LED in Figure 9.14 but other loads, such as a relay, may be substituted.

FET circuits

The action of FETs bears some similarity to the action of bipolar junction transistors and there are three equivalent connections that are used. For example, the **common source connection**, the equivalent of the common emitter connection, is used for switches and amplifiers. The **common drain connection** is used to build a **source follower** amplifier.

10

More semiconducting
devices

Most of the individual semiconductor devices found in electronic circuits are diodes or transistors as described in Chapters 7 and 8. There are many possible ways of assembling useful devices from semiconductors and this chapter describes some of them.

Unijunction transistor

This consists of a bar of n-type semiconductor with a region of p-type material near one end (Figure 10.1). The p-type material is manufactured by placing a piece of aluminium rod against the n-type bar and heating them in a furnace. Some of the aluminium diffuses into the p-type, converting it into n-type. The ends of the rod are connected to terminals known as base 1 and base 2. The emitter terminal is joined to the aluminium rod. When connected in a circuit, base 2 of the transistor is made more positive than base 1. There is a potential gradient along the bar, and a small current flows along it. The potential in the region of the p-type material is at a potential intermediate between that of base 1 and that of base 2. When the potential applied to the emitter is lower than this, the p-n junction is reverse-biased, and no current flows across it. If the potential applied to the emitter is gradually raised, there comes a point (the

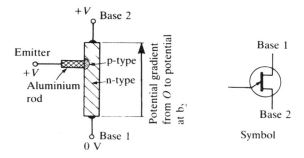

Figure 10.1 *The structure of a unijunction transistor*

99

peak point) at which the junction becomes forward-biased. Current flows from the emitter to base 1. The flow of current from the emitter sweeps holes into the base 1 region, reducing its resistance. This means that the emitter current is large, up to 2 A in a typical UJT. Although UJTs are not widely used, they are often used for building simple oscillators (p. 148).

Silicon controlled switch

Members of this family of devices are variously known as silicon controlled switches (SCSs), silicon controlled rectifiers (SCRs) or thyristors. These are four-layer devices (Figure 10.2a), consisting of alternate n-type and p-type layers. It is easier to understand their working if we think of the middle two layers being cut in halves, yet electrically connected (Figure 10.2b). Then the device becomes two connected bipolar transistors, one of the n-p-n type and one of the p-n-p type (Figure 10.2c).

The device is connected with the anode (the emitter of the p-n-p transistor) more positive than the cathode (the emitter of the n-p-n transistor). The third connection is to the base of the n-p-n transistor and the collector of the p-n-p

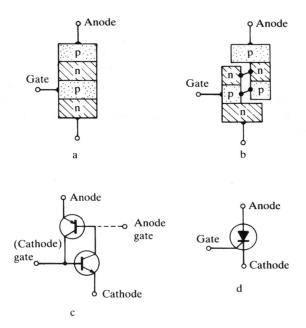

Figure 10.2 *Thyristors (silicon controlled switches) (a) Structure (b) The equivalent represented at two connected transistors (c) Schematic diagram of (b) (d) The symbol*

transistor, and is named simply the gate, or the cathode gate. To begin the cycle of operation the gate is at, or close to, the potential of the cathode, so the n-p-n transistor is switched off. No current is being drawn from the base of the p-n-p, so this is switched off too.

---

p-n-p transistor action

In the diagram of a p-n-p transistor, the arrowed line is the emitter. This is usually connected to the positive rail of the circuit. No current flows as long as the gate is held at, or close to, the same positive potential. But if the gate potential drops 0.6 V or more below the positive potential, current flows out of the gate, causing a current to flow through the transistor and out of the collector.

The n-p-n transistor is turned on when a short positive pulse is applied to the gate. This makes the potential at the collector of the n-p-n transistor fall, drawing current from the base of the p-n-p transistor. This turns the p-n-p transistor on (see box). Current flows from its collector. Even though the pulse originally applied to the gate may have finished, the current from the p-n-p transistor keeps the n-p-n transistor switched on. As a result current continues to flow through the device for as long as there is a positive potential at the anode. The device has been switched on by a very brief pulse at its gate and, once switched on, remains on.

If there is a break in the circuit to prevent current flowing to the anode, both transistors are switched off. They cannot be turned on again by reconnecting the supply to the anode. The only way to turn them on is to apply another positive pulse to the gate. The device may also be turned off by reducing the anode-cathode p.d. This reduces the current through the device, but it remains on until the current reaches a minimum value, known as the **holding current**. The device is also turned off by reversing the anode-cathode p.d., making the cathode positive with respect to the anode. In this way the device acts as a diode, allowing only one-way current, as its symbol (Figure 10.2d) suggests. But, unlike a diode, it does not turn on again when the p.d. is restored to its original direction.

Some devices are supplied with a fourth terminal, the anode gate. A positive pulse to this switches off the p-n-p transistor, cutting off the current even though the anode remains positive to the cathode. These devices are used wherever parts of a circuit are to be switched on or off under electronic control. Low-power versions usually go by the name of silicon controlled switch and may be used for switching indicator lamps in displays. Robust high-power versions, with provision for bolting to heat-sinks, are usually called thyristors and are used for power control, for example lamp-dimmers and motor speed controls (p. 90).

Triac

This is virtually a two-way thyristor, capable of conducting in either direction. Its structural diagram (Figure 10.3a) shows it to consist of two thyristors connected in parallel in the opposite sense and sharing a common gate. A positive pulse to the gate causes it to conduct. Like a thyristor, it switches off if the p.d. is reversed but, unlike a thyristor, it can be triggered into conducting in the opposite direction by another positive pulse to its gate.

Diac

The symbol of this device (Figure 10.3c) indicates that it has the action of a triac, but without the gate. It can be thought of as two diodes connected in parallel, in opposite sense. This gives it its alternative name, bi-directional trigger diode. The diac does not conduct if the p.d. between its terminals is less than a given amount, the breakdown voltage. The diodes are of the avalanche type (p. 72) so that current flows if the breakdown voltage is exceeded. An application of diacs appears on p. 90.

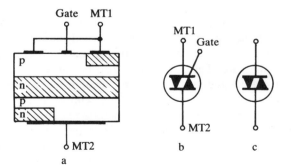

Figure 10.3   *Power-control semiconductor devices (a) A triac (b) Symbol for a triac (c) Symbol for a diac*

11

Sensors and transducers

Although the words 'sensor' and 'transducer' are often used as if they have the same meaning, their purpose is entirely different. A **sensor** is a device intended for detecting or measuring a physical quantity such as light, sound, pressure, temperature or magnetic field strength. It has an electrical output (often a varying potential or a current) which varies according to variations in the quantity it is detecting. A **transducer** is simply a device for converting one form of energy into another form of energy.

Some sensors are also transducers. For example, a crystal microphone (p. 107) converts the energy of sound waves into electrical potentials. It conveniently converts the sound energy into a form that can be handled by an electronic circuit. But some other sensors are not transducers. For example, a platinum resistance thermometer detects change in temperature as a change in the electrical resistance of its sensing element. Resistance is not a form of energy. Conversely, there are many transducers that are not sensors; examples include electric lamps (convert electrical energy to light energy), electric motors (convert electrical energy to motion), cells (convert chemical energy to electrical energy).

Descriptions of transducers begin on p. 115. Light sensors and transducers are dealt with separately in Chapter 12.

Temperature sensors

The resistance of metallic conductors increases with increasing temperature. If we measure the resistance of a length of wire, of known length, diameter and composition, we can determine its temperature. A **platinum resistance thermometer** consists of a coil of platinum wire wound on a ceramic former. One of its advantages is that it can be used over a very wide range of temperatures, from $-100°C$ to several hundred degrees Celsius. Unfortunately, the resistance of platinum, like that of all metals is very low, so that a long wire is needed for the coil, which is bulky. This means that it takes a relatively long time to acquire the temperature of the surroundings, and also that it is unsuitable for use in measuring the temperature in confined spaces. Finally, the change in resistance is slight and it requires special circuits (p. 60) to measure it. The platinum thermometer complete with measuring circuits is an expensive piece of equipment. It is chiefly used for high-precision measurements in industrial and scientific applications.

A **thermistor**, or thermally sensitive resistor, is much less expensive and easier to use. It consists of a sintered mixture of sulphides, selenides or oxides of nickel, manganese, copper, cobalt, iron or uranium. It is formed into a rod or bead or a disc, and may be encapsulated in glass (Figure 11.1). The resistance of a thermistor decreases with temperature (we say that it has **negative temperature coefficient**, ntc) and the change is relatively greater than for metals. This makes it easier to detect and measure small temperature changes. The small size of thermistors makes them suitable for measuring the temperature inside small spaces. For example, they have been used to measure temperatures inside the leaf of a plant. Their small size also means that they have very little effect on the temperature of the objects or spaces into which they are inserted, which makes their measurements more accurate. But thermistors have some disadvantages too. One is that a current must be passed through the thermistor in order to measure its resistance. This inevitably generates a certain amount of heat (p. 201), which leads to errors in the readings. Measuring circuits must be designed to minimize the current through the thermistor. The other problem is that the response of a thermistor is not linear. In other words, when we plot resistance against temperature we do not obtain a straight line (Figure 11.1). This is not significant over a small temperature range but reduces precision over a wider range.

Thermistors are also made with **positive temperature coefficient** (ptc). This means that their resistance increases with increasing temperature. Such thermistors are often used for temperature compensation in electronic circuits.

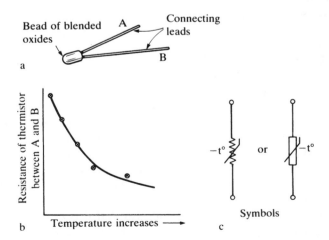

Figure 11.1 *Thermistors (a) Typical thermistor (b) How its resistance changes with temperature (c) The symbols*

Currents in parts of a circuit may increase to dangerously high levels if the temperature of the circuit becomes unduly high (see Thermal runaway, p. 65). If a ptc thermistor is included in the circuit, its increased resistance reduces the current.

A **bandgap sensor** depends upon two opposite changes of p.d. The base-emitter p.d. of a transistor decreases with increase of temperature. It is also possible to connect two transistors so that they produce a p.d. that increases with temperature. By suitable choice of resistors it is possible to design a circuit in which these counter-balancing p.d.s result in a p.d. that increases by a precisely determined amount as temperature changes. Bandgap sensors such as these are made as integrated circuits, but sealed in a three-wire plastic package like that of an ordinary transistor. The small package can be inserted in small spaces, has little effect on their temperature, and responds rapidly to changes in temperature.

Connected as shown in Figure 11.2, the output of the sensor ranges from 0 V to 1.1 V as the temperature ranges from 0°C to 110°C. The response is linear, so that if, for example, we find that the output is 0.37 V, we know that the temperature is 37°C. The output is measured either by connecting a voltmeter to the circuit as in the figure, or feeding the output to a more complicated measuring and recording circuit, or indirectly to a computer.

The sensor of the passive infra-red detectors, used to trigger a floodlight or siren when an intruder comes into their field of view, is a **pyroelectric device**. The device consists of a crystal of a substance such as lead-zirconate-titanate which has been heated and then cooled in a strong magnetic field. It is sensitive to changes in the amount of infra-red falling on it. Any slight change causes a p.d. to appear between opposite faces of the crystal. This p.d. can be measured and used to trigger alarms. When used in a security system, the sensor is placed behind a plastic sheet which has a raised pattern formed on it. The sheet acts as a kind of lens, focusing radiation from the surveyed area on to the sensor. But

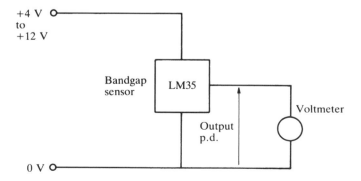

Figure 11.2 *Using a band-gap sensor to measure temperature*

the lens has the effect of dividing the area into radiating zones. Alternate zones are visible and invisible to the sensor. When a person or any object warmer than the surroundings moves within the surveyed area it inevitably moves from a visible zone to an invisible zone, or the other way about. This sudden change in the amount of radiation received by the sensor, causes it to trigger the alarm system.

This type of sensor is also used in industrial temperature measuring instruments (p. 283)

A **thermocouple** relies on the Seebeck effect to measure differences of temperature. It consists of two junctions (Figure 11.3) between dissimilar metals, such as iron and constantan (a copper-nickel alloy). The junctions can be made by simply twisting the ends of the wires together, but it is more reliable to solder them together. The junctions are subjected to two different temperatures. One, usually called the cold junction, is placed where is it subject to a steady reference temperature (perhaps in the case of the instrument). The other junction, the hot junction, is subject to the temperature being measured. Under these conditions a p.d. is generated across the pair of junctions and is proportional to their temperature difference. This p.d. is measured by a circuit which produces a reading on a display. A thermocouple is generally used for measuring very high temperatures of several hundred or even several thousands of degrees Celsius. The precise temperature of the cold junction within a few tens of degrees is not significant. For high temperatures, the junctions are usually metals that can withstand high temperature without corroding. An example is a junction between platinum and a rhodium/platinum alloy. A **thermopile** is a device in which several thermocouples are connected in series with their cold junctions grouped close together, and their hot junctions grouped likewise. This produces more current than a single thermocouple, so can be used for measuring smaller temperature differences.

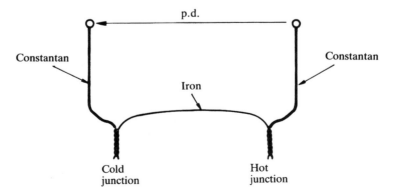

Figure 11.3 *The principle of a thermocouple*

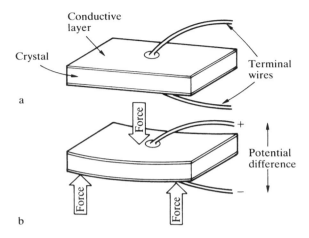

Figure 11.4 *The piezo-electric effect (a) The transducer (b) An applied force generates a p.d. or applied p.d. generates a force*

Sound sensors

A useful property of crystals of quartz is that when they are slightly stressed (bent out of shape) a p.d. is developed between opposite sides of the crystal (Figure 11.4). This is known as the **piezo-electric effect**. The p.d. produced is very small but it can be amplified to produce a usable electrical signal. In a piezo-electric microphone, the sound waves impinge on a crystal or disc of piezo-electric material. They produce changes in pressure which vibrate the crystal, making it change its shape according to the shape of the sound waves. A varying p.d., corresponding to the sound waves appears at the terminals of the microphone.

Piezo-transducers

Transduction by a quartz crystal operates in both directions. In the crystal microphone, the crystal is stressed and it generates a p.d.. Sound (mechanical) energy is converted into electrical energy. Conversely, if a p.d. is applied to opposite faces of the crystal, it changes shape. Electrical energy is converted into mechanical energy. This effect is made use of in the small **sounders** used to produce bleeps and other sounds from digital watches and various electronic gadgets. The principle is also used in time-keeping itself (p. 155).

Another type of microphone relies on electromagnetism. There are two kinds of these, **moving-iron**, and **moving-coil**. In the moving-iron microphone (Figure 11.5), the sound waves vibrate the iron diaphragm and cause fluctuations in the magnetic field strength between the diaphragm and the poles of the magnet. The nearer the diaphragm is to the poles, the stronger the field. The changes in field strength induce currents in the coils, producing a p.d. across their terminals. The moving-coil microphone works like a loudspeaker in reverse (p. 115). In fact, in simple intercom systems, the loudspeakers may double up as microphones.

A **capacitor microphone** has two plates separated by an air gap. One plate is a rigid metal disc and the other is a flexible metal disc or a disc of metallized plastic. Sound vibrates the thin disc altering the spacing and hence the capacitance between the two discs. The microphone requires a source of e.m.f. to charge it, but has the advantage that the sound quality is very high. A relative of this type of microphone is the **electret**. This has a dielectric material between the plates. During manufacture the dielectric is heated and then cooled while it is being subjected to a strong electric field. This causes a permanent electric field to be held in the dielectric. Vibration of the electret causes a p.d. to be generated between the two plates. Both types of capacitor microphone produce only a small signal and their output impedance (p. 170) is very high. For this reason a special pre-amplifier is usually included inside the microphone case.

A ribbon microphone consists of a narrow ribbon of aluminium suspended between the poles of a magnet. The ends of the ribbon are connected to an amplifying circuit. Arriving sound waves make the ribbon vibrate and this induces alternating currents in the ribbon. The current signal is amplified. Unlike the other microphones, this type does not respond to the changes in pressure caused by sound but to the velocity of the air particles. Being a very light ribbon it has low inertia so responds very faithfully to the sound waves,

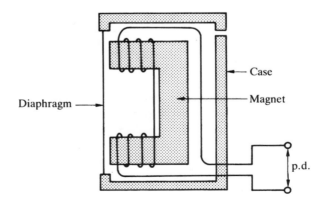

Figure 11.5   *A section through a moving-iron microphone*

and its sensitivity to high frequencies is good. The main drawbacks are that a ribbon microphone is expensive and requires careful handling to avoid damage.

Mechanical sensors

A **strain gauge** consists of a thin metal foil with a zig-zag pattern of parallel constructors. The foil is mounted on a plastic backing and cemented to the object (such as a steel beam) that is under strain, with the conductors parallel to the direction of greatest stress. When a load is applied to the object, so that it becomes under strain, the conductors become stressed in the same way. They are made longer and therefore become thinner, and their resistance increases. By measuring their resistance (generally with a bridge, p. 60) we are able to determine the amount of strain.

Other forms of strain gauge rely on the piezo-electric effect (p. 107). The sensor is a strip of piezo-electric material cemented to the object under strain. When the object is strained, distortion of the material causes a p.d. to develop between its opposite faces. This is measured by a suitable electronic circuit.

The simplest form of motion detector is a **mercury switch**. It is also a position or vibration detector. It consists of a metal capsule into which are sealed two contact wires and a quantity of mercury. When the capsule is upright and stationary the mercury settles in the bottom of the capsule and makes an electrical contact between the two wires. If the capsule is tilted, moved rapidly, or vibrated, the contact is momentarily broken. The break in contact is detected by a suitable electronic circuit.

The basis of many types of position detector is the **linear variable differential transformer** (Figure 11.6), or LVDT. An alternating current is passed through

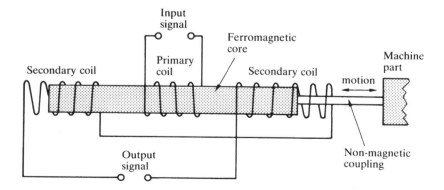

Figure 11.6 *The principle of an LVDT*

the central primary coil. It generates an alternating magnetic field which is linked to the two secondary coils by the ferrite core. The secondary coils are connected in series and the direction in which they are wound is such that, when the core projects into each coil by equal amounts, the induced alternating e.m.f.s are equal and opposite. They cancel out and the total p.d. across the two coils is zero. The core is mechanically attached to the machine part whose position is to be registered. As the part moves, the core moves so that it extends more into one secondary coil than the other. The e.m.f.s are then unequal and the amplitude of the induced current increases. By measuring the amplitude of the signal it is possible to determine the position of the machine part.

Acceleration may be measured by a device like that shown in Figure 11.7. A disc of piezo-electric material (p. 107) has a heavy metal mass attached to it. When the sensor is subjected to acceleration in an upward or downward direction, the material is either compressed or expanded. This causes a p.d. to develop between opposite surfaces of the disc. The p.d. is measured by an external circuit and the result expressed in terms of acceleration.

The piezo-electric effect may also be used for measuring gas pressure, including atmospheric pressure. The essential part of a **pressure sensor** is a chamber divided into two parts by a thin diaphragm. Very often one part of the chamber is open to the atmosphere and the other part has a port by which it is connected to the vessel in which pressure is to be measured. The difference in pressure between the two parts of the chamber cause the diaphragm to bulge one way or the other. To measure the pressure difference we measure the amount of bulging. If a crystal of piezo-electric material is mounted on the diaphragm, bulging of the diaphragm causes distortion of the crystal, which produces a measurable p.d. The p.d. is proportional to the pressure difference. The operation of this sensor is very similar to that of a crystal microphone which, after all, is sensitive to the pressure differences produced by sound waves.

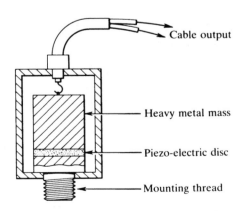

Figure 11.7 *A piezo-electric transistor used as an accelerometer in missile research*

In **capacitative pressure sensors** a metal plate is mounted on the diaphragm and forms one plate of a capacitor. The other plate is mounted parallel with it in a fixed position. Movements of the diaphragm cause the capacitance to vary, and the variations are measured by a suitable circuit.

Another way of measuring the distortion of the diaphragm is to mount a strain gauge on it (p. 109). In the most recent pressure sensors, the diaphragm is a wafer of silicon. Using techniques similar to those used in manufacturing integrated circuits (p. 136), a piezo-electric strain gauge is built up on the wafer, complete with a bridge circuit and amplifiers. The circuit may include a thermistor (p. 104) to compensate for variation in output due to temperature changes. The sensor is housed in a case suitable for mounting directly on a printed circuit board (p. 134) and output p.d. is easily convertible to pressure readings.

Piezo-resistive pressure sensors may be inserted in the combustion chamber of an internal combustion engine to detect irregular pressure changes due to misfiring of the engine. The data is fed to an in-vehicle computer for analysis. The aim is to run the engine with as high an air:fuel ratio as possible, so as to reduce the amount of pollutants (carbon dioxide and nitrogen oxide) in the exhaust, as well as to reduce fuel consumption. If the ratio is too high, the engine becomes unstable and misfires, but this condition is detected by the pressure sensor and the proportion of air is reduced until the engine runs smoothly again.

Magnetic field sensors

If a bar of semiconducting material is placed in a magnetic field (Figure 11.8), the charge carriers are deflected. The effect is to produce a p.d. across the semiconductor. This is known as the **Hall effect**. A transistor or other amplifying device is connected so as to detect the p.d.. A Hall-effect detector incorporates the semiconductor and amplifier in a three-terminal package which looks just like an ordinary transistor. Usually the sensor and amplifying circuit are integrated on the same chip. Detectors may have different types of response. Many act as on-off switches, which have the advantage of being bounce-free (p. 199). The potential at their output terminal swings abruptly from 0 V to the supply voltage when the magnetic field changes. Some of these respond to the field direction; a field in one direction turns them on, then they stay on until they are subjected to a field in the opposite direction. Others are sensitive to field strength; they are turned on when the field exceeds a certain strength. They stay on until the field strength falls below a lower level.

Switching Hall-effect devices have many applications as position sensors. If a moving part of a machine has a magnet attached to it, its position can be monitored by Hall-effect devices located on the frame of the machine. They are also used in tachometers, for measuring the rate of revolution of an axle or

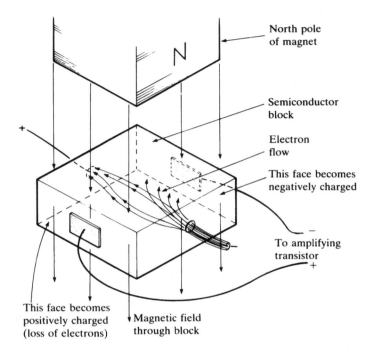

North pole
of magnet

Semiconductor
block

Electron
flow

This face becomes
negatively charged

To amplifying
transistor

This face becomes
positively charged
(loss of electrons)

Magnetic field
through block

Figure 11.8   *The Hall effect, showing how a magnetic field causes the electrons flowing through the device to be deflected*

wheel. If a notch is cut in a rotating ferrous metal part of the machine, a Hall effect device is affected when the notch passes it, so giving an alternating signal that can be used to measure the rate of rotation. A permanent magnet placed behind the device is used to increase the local field strength, making the device more sensitive to the rotating notch. A device such as this, which measures the rate of rotation, is known as a **tachometer**.

Other types of Hall-effect device have a linear output. The output voltage is proportional to the field strength. This type can be used for precise position measurement of a moving part of a machine which has a pair of magnets attached to it (Figure 11.9a). Linear Hall-effect devices are used to measure large currents. In Figure 11.9b the current (a.c. or d.c.) passing along the cable generates a magnetic field proportional to its strength. The magnetic field is confined to the toroid of ferrite. A Hall-effect device positioned in the gap in the toroid has an output proportional to the field strength and hence proportional to the current in the cable.

Magneto-resistive sensors are described on p. 228.

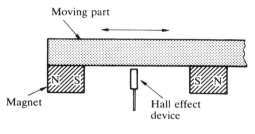

a

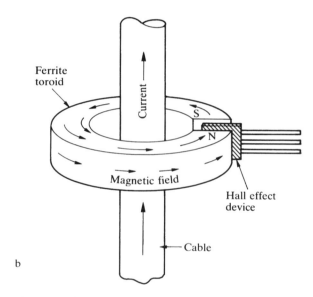

b

Figure 11.9   *Applications of Hall-effect devices (a) Precision position measurement (b) Current measurement*

Humidity sensors

One form of humidity sensor is based on a film of hygroscopic material, which absorbs water in humid conditions and loses it in dry conditions. At any given humidity the amount of water present in the material reaches an equilibrium. This affects the resistance between two metallic electrodes (Figure 11.10). The resistance is measured by a circuit which then displays the result as a reading of humidity. In another type of sensor, the hygroscopic material forms the dielectric of a capacitor. Changes in humidity result in measurable changes in capacitance.

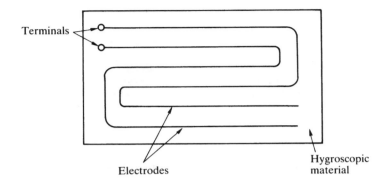

Figure 11.10   *A resistive hygrometric (humidity) sensor, seen in surface view*

pH sensor

A pH meter measures the degree of acidity or alkalinity of a solution. The pH scale runs from pH0 for a very strong acid to pH14 for a very strong alkali. Water, which is neutral (neither acidic or alkaline) has a pH of 7. Two platinum-wire electrodes are immersed in the test solution. One electrode is surrounded by a standard solution containing dissolved salts. The electrode and solution are separated from the other electrode by a glass bulb which allows hydrogen ions to pass through it but restricts the movement of other ions. In effect, this is a cell in which the p.d. developed between the electrodes is related to the hydrogen ion concentration in the solution. This is directly related to the pH. A circuit measures the p.d. between the electrodes and displays the pH on a scale.

Gas sensors

Gas sensors are used for detecting combustible gases. The sensing element is a platinum wire coated with a mixture of substances which catalyze the oxidation of combustible gases. The coil is heated by passing a current through it; this causes combustion to occur, generating additional heat and thus increasing the resistance of the coil. The circuit also includes a dummy element from which the catalysts are absent. This too is heated by passing a current through it but does not gain the additional heat from gas combustion, so its resistance is not raised. A bridge circuit (p. 60) compares the resistances of the two coils. When gas is present, the bridge goes out of balance and this may be used to trigger an alarm. In practice the two coils are contained within a dome of stainless steel mesh. This is to prevent combustion around the sensing element from spreading to the atmosphere outside, possibly causing an explosion. Gases that are

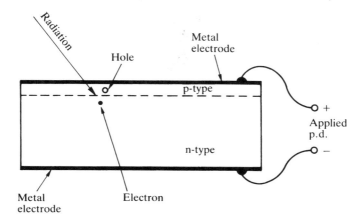

Figure 11.11   *An extrinsic semiconductor nuclear radiation detector, seen in section*

detectable with gas sensors include methane (present in town gas and natural gas), hydrogen, and alcohol vapour.

Nuclear radiation detectors

A wide range of radiation detectors has been invented, including the ionization chamber, the scintillation counter and the Geiger-Muller tube, but we describe a more recent solid-state device which makes use of the properties of semiconductors. The extrinsic semiconductor nuclear radiation sensor (Figure 11.11) consists of a slice of n-type silicon, the surface layer of which is doped to convert it to p-type. The upper and lower surfaces of the slice are metallized to form electrodes. In operation, an external circuit produces a p.d. across the device. Nuclear radiation passing into the p-type layer displaces electrons from the atoms, creating electron-hole pairs. These are accelerated by the electric field within the slice, creating further electron-hole pairs. The result of the arrival of nuclear radiation is thus a sudden increase in the number of charge carriers and therefore a pulse of current through the device. The pulse is detected by the external circuit. The rate at which pulses occur is proportional to the amount of radiation being received.

Sound-producing transducers

Most loudspeakers work by electromagnetism. A powerful magnetic field is produced by a specially-shaped permanent magnet (Figure 11.12), rigidly

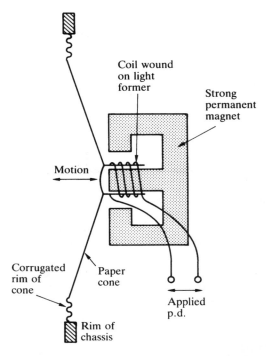

Figure 11.12   *A section through an electromagnetic loudspeaker*

attached to the chassis of the speaker. A lot of research has gone into developing magnetic materials that can provide a high-intensity field, and are not readily demagnetized. The voice coil is a coil of thin wire usually wound on a light plastic core. It is attached to the rear of a paper or plastic (mylar) cone. The code and coil are supported by a ring of corrugated paper or plastic, which allows the cone to vibrate backward and forward with great freedom. When a signal current from an amplifier is passed through the coil, a varying magnetic field is generated within the coil. The interaction between the field of the permanent magnet and that of the coil, forces the coil to vibrate. These vibrations are transferred to the cone which, in turn, makes the air in front of the cone vibrate, generating sound waves in the air.

The impedance of a loudspeaker coil is generally quoted for a frequency of 1 kHz (p. 38). Most loudspeakers have low impedance, 4 Ω or 8 Ω being common values. The low impedance allows a large current to flow, so that the speaker has a high power rating (p. 26) and the sound produced is loud. Small speakers of higher impedance (often 64 Ω) are sometimes used in battery-powered radio sets and tape players.

The other common way of converting electrical energy into sound is by a piezo-electric **sounder** (p. 107). Electronic sirens and buzzers have a sounder of this type, with built-in transistor circuits to generate the alternating p.d. required.

Motion-producing transducers

Probably the most familiar example of a device which converts electrical energy into motion is the **electric motor**. The operation of electric motors is more a concern of electrical engineering rather than of electronics, so we will not describe these here. The simplest motion producer is a **solenoid**, which consists of a coil with a soft iron core (or armature) which slips easily in and out of the coil (Figure 11.13). Soft iron is used because it does not retain its magnetism when the current is switched off; in other words, it does not become permanently magnetized. The armature begins with only a part of it within the coil. When a current is passed through the coil, the armature is drawn into the coil with great force. This gives solenoids many uses in driving parts of robot machines, or operating bolts on doors and windows. Turning off the current does not force the armature out again, so the reverse action must be provided for by some external mechanism. This may be simply a spring, compressed when the solenoid is energized and released when the current is cut off. In robots the reverse action may be effected by a second solenoid operating in the reverse direction.

A **relay** is an electromagnetically operated switch. It consists of a coil wound on a soft-iron core, and a soft-iron armature mounted on a pivoted lever (Figure 11.14). The lever has contacts on it which close against other stationary contacts mounted on the chassis of the relay. The arrangement of contacts depends upon the model of the relay. Usually there is a pair of change-over contacts as shown in the figure. When there is no p.d. across the coil, the

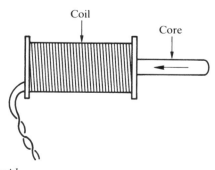

Figure 11.13 *A solenoid*

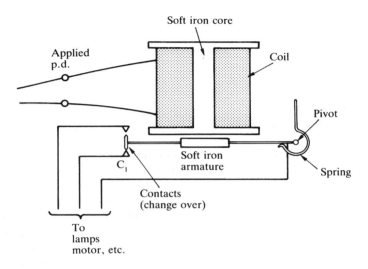

Figure 11.14   *How a relay works*

movable contact closes against the lower fixed contact. This pair of contacts is described as being **normally closed**. There is an open circuit between the movable contact and the upper fixed contact. It is described as **normally open**. The situation is reversed when a p.d. is applied to the coil. The movable contact is moved away from the lower fixed contact and closes against the upper contact.

Relays are normally used for controlling lamps, motors and similar devices. A small current, perhaps that provided by a transistor switch (p. 75) or the output from a logic gate may be used to control a very much larger current. The changeover contacts make it possible to turn one device on when another is turned off. Some relays have a more complicated array of contacts, allowing several devices to be switched on or off simultaneously. Although junction transistors and MOSFETs can also be used for this purpose, a relay has the advantage that it can have contacts suitable for operating at high voltages, with high currents. A particular advantage is that relays can switch alternating currents (such as the mains), which is not possible with transistors.

Relays are made in a number of sizes, including miniature types suitable for mounting on a printed circuit board (p. 134). Some of these are contained in a package identical to that used for integrated circuits (p. 138).

A **reed switch** has the same function as a relay. Reed switches are not able to carry such heavy currents as a typical relay, but they act more quickly. The switch consists of two springy contacts sealed in a capsule to exclude dust and dampness. The ends of the contacts overlap, as shown in Figure 11.15, but do not touch. In one type of reed switch, there is a coil wound round the capsule. When a current is passed through the coil the contacts become magnetized

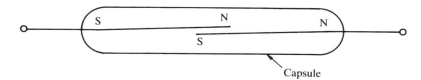

Capsule

Figure 11.15   *A reed switch*

temporarily. They are both magnetized with the same polarity so that the north end of one contact is close to the south end of the other. They are attracted toward each other and meet, completing an electric circuit. When the current to the coil is turned off, they lose their magnetism and spring apart.

A reed switch (without a coil) may also be operated by bringing a permanent magnet close to it. This is useful for monitoring the action of machinery; the magnet is attached to a moving part and one or more reed switches are fixed where they can detect the position of the magnet. Reed switches are used in security systems to detect if doors and windows have been opened. To protect a door, a reed switch is mounted in or on the door frame. A permanent magnet is mounted in or on the door itself so that, when the door is shut, the magnet comes close to the switch. When the door is opened, even by as little as a centimetre, the magnet is separated from the switch and the contacts spring open. A number of such switches on the doors and windows of a building may be connected in series to form a loop. If all doors and windows are shut, all switches are closed and current passes round the loop. But if any one door or window is open, its switch contacts open too, breaking the loop. This breakage is detected by a centrally-located circuit, which immediately triggers an alarm.

12

Optoelectronic sensors

There are so many devices that are sensitive to light, and so many applications for them, that we are devoting a separate chapter to the optoelectronic sensors.

Light-dependent resistor

This consists of a disc or square block of semiconducting material, usually cadmium sulphide, with two electrodes printed on its surface (Figure 12.1) The electrodes are shaped so that there is a narrow gap between them, but their edges are waved to make the length of the gap as great as possible. The whole is encapsulated in clear plastic. When a p.d. is applied to the electrodes a current passes from one electrode to the other. Under dark conditions the current is limited by the small amount of charge carriers present in the semiconductor. The resistance of the device is high, perhaps as great as $10\,M\Omega$. When light falls on the disc, its energy excites electrons in the atoms of the semiconductor and many electron-hole pairs are generated. This greatly increases the number of charge carriers and the current increases. Putting it another way, the resistance decreases. In bright light it may fall as low as $100\,\Omega$.

Because the resistance of the device depends on the intensity of light falling on it, it is known as a **light-dependent resistor**, often shortened to LDR. Another name for this device is **photo-conductive cell** (PCC) but, unlike other cells, it does not generate a p.d., so this name is misleading.

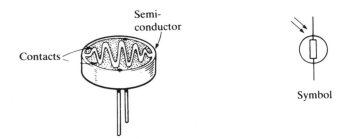

Figure 12.1 *A light-dependent resistor and its symbol*

120

LDRs have many applications for measuring light levels, for example, in the exposure controlling circuit of a camera. They operate in a wide range of light intensities. Their main disadvantage is that their response time is low. An LDR takes about 75 milliseconds to respond to a fall of light intensity (the time required to sweep the existing charge carriers away into the external circuit) and they may take several hundred milliseconds to respond to a rise in intensity. This is no disadvantage in some applications but in others, such as reading a bar-code (p. 276) their response is too slow.

Photodiode

The p-n junction of an ordinary diode is sensitive to light. Usually, light is excluded by sealing the diode in a light-proof package of plastic or painted glass. A photodiode has a transparent cover; it may either be a glass capsule or a metal can with a lens at its end to focus light on to the p-n junction. Infra-red photodiodes have a package that is opaque to visible light but transparent to infra-red.

The effect of light is to excite the electrons in the atoms of the semiconductor, causing some of them to escape, leaving holes. The increase in electrons and holes means that there are more charge carriers and, for a given applied p.d., the current increases. In most circuits the photodiode is reverse biased, so that charge is carried by extrinsic charge carriers. The response time of a photodiode is typically 250 nanoseconds. Compare this with the response time of an LDR, measured in milliseconds, a unit a million times longer. Photodiodes, particularly the infra-red variety, are used in security systems and in TV remote control systems.

PIN photodiodes have a layer of intrinsic (not doped) semiconductor between the p-type layer and the n-type layer. The effect of the intrinsic layer is to reduce the capacitance between the p-type and n-type layers (p. 33 and p. 73). This makes it easier for the diode to respond more rapidly to changes in light levels. Response times of the order of 0.5 nanoseconds are typical. This is important in diodes being used to receive high-frequency signals by optical fibre.

Phototransistor

This has a similar structure to an ordinary n-p-n junction transistor, except that the case has a glass window, usually shaped like a lens to focus light on to the transistor (Figure 12.2). The cheaper types are encapsulated in transparent plastic. As the symbol indicates, a phototransistor does not necessarily have a connection to its base layer. When light falls on the transistor, atoms in the

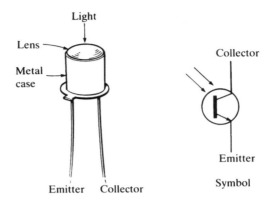

Figure 12.2   *A phototransistor package and its symbol; some types, such as the one illustrated here, have no connection to the base*

collector region become excited and emit free electrons. This creates electron-hole pairs. These electrons leak through the reverse-biased collector-base junction. Once in the base region they pass on to the emitter; they are the equivalent of a base current. This base current is amplified by the usual transistor action (p. 83), resulting in an increase in the collector current. Some types of photo-transistor have a connection to the base layer. This makes it possible to bias the transistor (p. 168) so that its response time is reduced.

The amplification due to transistor action means that phototransistors are more sensitive than photodiodes. But their response time is much slower, usually several milliseconds. With the development of integrated circuit techniques, it is now possible to manufacture a photodiode and an amplifier circuit on the same chip. This combines high sensitivity with a short response time, so that phototransistors are less often used today than formerly.

Camera on a chip

The ultimate in optoelectronic sensors is a complete camera on a single chip. The WL1070 has a 160 × 160 optical sensor array on to which an image is focused by a built-in plastic lens. The chip also includes analogue-to-digital converters to record brightness and all the control and logic circuits needed to produce a digitized version on the image. The integrated circuit takes only 30 mA so it is suitable for use in hand-held battery-powered equipment.

Photovoltaic cell

Unlike the other devices described so far in this chapter a photovoltaic cell generates a p.d. when light falls on it. This makes it a transducer as well as a sensor (p. 103). The early photovoltaic cells consisted of an iron plate coated on one side with selenium (a semiconductor), and were known as **selenium cells**. They were used in early motion picture projectors, to read the optical soundtrack. Nowadays photovoltaic cells are based on silicon as the semiconductor. A typical p-on-n photovoltaic cell is shown in Figure 12.3. A very thin layer of p-type is produced on one surface of a block of n-type silicon. The p-type layer is so thin that light penetrates it and reaches the depletion region. As explained on p. 69, there is a naturally-occurring p.d. of about 0.6 V across this junction, known as a **virtual cell**. The n-type layer is positive with respect to the p-type layer.

   When light falls on the cell, the atoms are excited and liberate electrons, creating electron-hole pairs. The electrons are attracted toward the positive n-type layer by the field of the virtual cell. Conversely, the holes are attracted toward the p-type layer. If the two layers are connected through an external circuit, the extra electrons in the n-type later pass through the circuit to combine with the extra holes in the p-type layer. In other words there is a current in the external circuit. This continues to flow for as long as light is falling on the cell and renewing the supply of electrons and holes. The p.d. across the cell is about 0.6 V, due to the virtual cell but, now that there is a source of energy (light), the virtual cell is able to supply a continuous current.

   Photovoltaic cells are made with a relatively large area to produce a reasonably large current. Response time is short (typically 50 ns) and response is

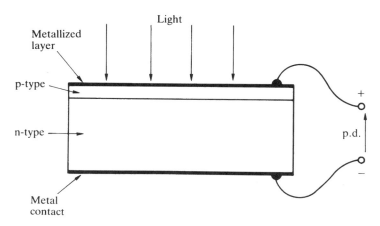

Figure 12.3   *A section through a silicon p-on-n photovoltaic cell*

linear. This makes them useful in optical instruments and for optical-fibre communications. One type is made that incorporates a colour filter in its lens system to give it a similar response to that of the human eye. This has applications in colour-matching instruments.

Solar cell

This has the same essential structure as a photovoltaic cell and works in the same way. The chief difference is that a solar cell has a very much larger area than an ordinary photovoltaic cell. Figure 12.4 shows an n-on-p solar cell. Instead of a continuous n-type layer the electrode consists of a pattern of strips. This arrangement allows more light to penetrate to the depletion layer, and offers less resistance to the flow of current to the terminal. Solar cells are usually connected in series to produce a battery with high total p.d. Batteries with an output p.d. up to 30 V can be purchased ready-made. Under full sunlight, these deliver a current of several hundred milliamps. Solar batteries are used to provide power for equipment that is remote from mains power supplies. Microwave link transmitters on mountain tops are a good example. They are also used for powering spacecraft and satellites. Experimental vehicles have been successfully powered by solar batteries, including an aeroplane that flew, using power from solar batteries located on its wings. Moon buggies are another example of vehicles powered in this way.

Unfortunately the efficiency of solar cells is low. They convert only 15% of the light energy falling on them. Research is being directed toward improving efficiency and reducing cost.

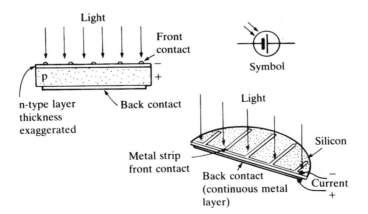

Figure 12.4   *A silicon photovoltaic cell*

Pyroelectric sensor

Pyroelectric sensors are particularly sensitive to infra-red light, so can also be classed as heat sensors. For this reason they are described in Chapter 11.

13
Light sources and displays

One of the most familiar of light-producing transducers is the filament lamp. Current passing through the filament causes it to glow white-hot. Some of the electrical energy is converted to heat, but much is converted to light. Although electronic circuits may sometimes include filament lamps as indicators, the light source most representative of electronics is the light-emitting diode.

Light-emitting diode

The most typical form of a light-emitting diode, or LED is illustrated in Figure 13.1. The case is made from transparent plastic, usually in the same colour as the LED, though clear plastic is used sometimes. The most common colour is red, since red LEDs produce more light than LEDs of other colours.

Energy is released when the electrons in the n-type material of a diode meet the holes in the p-type material. In a silicon or germanium diode this energy is lost as heat. If the material is doped with compounds containing gallium arsenide, the energy is released as visible light instead. The colour of the emission depends upon the exact composition of the dopant. Compounds based on gallium arsenide have electrons of relatively low energy and so emit radiation of the longer wavelengths, particularly infra-red and red, but also yellow and green. LEDs containing gallium nitride are also made for higher luminance. For some time it was not found possible to produce a blue LED, but blue

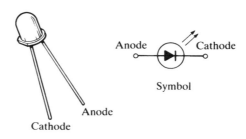

Figure 13.1 *A light-emitting diode*

LEDs are now available, based on silicon carbide or, for higher intensity, indium gallium nitride.

Infra-red LEDs are used to produce invisible beams in security systems; the beam is directed at a sensor, usually an infra-red photodiode. When the beam is broken by an intruder passing between the source and the sensor, the alarm is sounded. Infra-red LEDs are also used in TV and video player remote control systems. The infra-red LEDs in the hand-held controller are used to flash a coded series of pulses. The pulses are detected by an infra-red photodiode in the TV set and decoded. This causes a change of programme, increase of volume or other action depending on which button was pressed on the controller.

LEDs are made in a range of sizes from the miniature ones only 2 mm in diameter to the jumbo-sized LEDs which are 10 mm in diameter or more. They are also made in a variety of shapes: circular, triangular, rectangular. A row or 'stack' of LEDs placed close together is used to make a bar graph. These are often found on audio equipment to indicate sound volume. At low volume only the bottom few LEDs of the bar graph are lit, but as volume increases, more and more of the LEDs light up.

LEDs are used to display letters of the alphabet or numerals. One way of doing this is to arrange LEDs in a rectangular array of, say, seven rows of five LEDs. Different characters are produced by turning on the appropriate LEDs. A row of such arrays is used for displaying messages. If only numerals and a few letters are required, the seven-segment display (Figure 13.2) is the simplest and most popular. Seven rectangular LEDs, and usually one or two smaller square LEDs, are mounted in a plastic block. By turning on the appropriate segments any numeral can be displayed. The display may be controlled by a special integrated circuit which lights the correct segments when a coded number is sent to it. The small square is used as a decimal point.

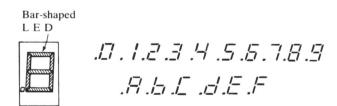

Figure 13.2 *A seven-segment light-emitting diode, showing the characters it can produce*

Optocoupler

One special application of LEDs is in opto-couplers (Figure 13.3). An LED and a phototransistor are sealed in a light-proof plastic package, so that light from the LED is received by the phototransistor. When the LED is turned on

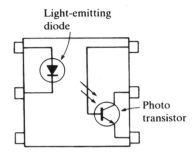

Figure 13.3 *An opto-coupler, consisting of an LED and a phototransistor packaged together*

by a current supplied from an external source, the phototransistor is turned on. If the phototransistor is wired as a switch, this can turn on other devices. The device acts as a relay but operates considerably faster. Such devices are used when signals must be sent from one circuit to another but it is not suitable for the circuits to be electrically connected. For example, the LED may be switched on and off by a logic circuit operating at low voltages, while the transistor may be part of another circuit in which high voltages are present. The two sides of the device are electrically insulated to withstand voltage differences of several thousands of volts.

Optocouplers are sometimes used for safety where one part of a circuit is to be connected to the human body, and the other part of the circuit operates at dangerously high voltage. As well as simple 'on-off' signals, the optocoupler can transmit continuously varying analogue signals.

Opto-coupled thyristors and triacs are used to switch alternating currents at mains voltages by means of low-voltage control circuits.

Laser

A laser is a device in which a large number of atoms are excited in such a way that they all emit radiation of a single wavelength in phase with each other. A laser produces a beam of light of very high intensity and with high energy content. In industry and in scientific research, lasers are used to produce high concentrations of energy in a small area, to generate extremely high temperatures and pressures. Because laser light is monochromatic (a single wavelength) it can be focused into a narrow pencil of light. Laser beams may travel for many kilometres without appreciable widening and laser beams have also been projected to the Moon and reflected back to Earth. The laser is not an electronic device so a description of how it works is outside the scope of this book.

Laser diode

A laser diode is similar to an LED in many respects but it produces a narrow beam of high intensity. The radiation produced by a laser diode (usually in the red or infra-red regions of the spectrum) is of a single wavelength so that, if it is passed through glass panels or lenses, the beam is refracted as a whole, not split into different wavelengths. When the light is produced by the diode element it spreads out over a wide angle, but is easily focused to form a narrow and virtually parallel-sided beam. The beam of a typical laser diode is 4 mm × 0.6 mm, widening only to 120 mm at a distance of 15 m. The ability to produce a very narrow beam makes the laser diode specially suitable for use in bar code readers (p. 276) and in CD and CD-ROM players (p. 213). Since a laser diode can be switched on and off at high frequencies, even as high as 1 GHz, they are also used for telecommunications by optical fibre (p. 235).

Liquid crystal display

A liquid crystal display consists of two glass plates with a thin layer of liquid crystal material sandwiched between them (Figure 13.4). The inner surfaces of the glass are printed with a transparent film of conductive material, such as indium oxide or tin oxide to form electrodes. The liquid crystal material is an organic substance that is liquid but, unlike most liquids, has crystalline properties. Its molecules are long and narrow in shape and the naturally-occurring forces between them tend to make them lie together in a regular pattern. This gives them their crystal-like properties. The most important effect of their crystalline nature is that they rotate the plane of polarization of light as it passes through the crystal. As Figure 13.4a shows, the plane of polarization is rotated by exactly a quarter of a turn if the film is of the correct thickness. If films of polarizing material are placed on either side of the display, with their polarizing directions at right-angles, light enters at the top, is polarized, its plane is rotated 90° by the crystal, and it is able to emerge through the bottom polarizer. Light apparently passes unaffected through the display.

If a p.d. is applied between the electrodes, the electric field alters the arrangement of the molecules (Figure 9.4b). There is no rotation in the crystal. Light passing in through the top polarizer is not able to pass out through the bottom polarizer. The liquid between the electrodes appears to be black. If the p.d. is removed the molecules immediately return to their crystalline arrangement and the display becomes transparent again.

In a liquid crystal display (LCD) the back electrode (or back plane) usually covers the whole inner surface. A pattern of separate electrodes in printed on the front inner surface. This takes the form of one or more seven-segment arrays (Figure 13.2) and a variety of other symbols, such as maths signs and legends, depending on the application. Very narrow tracks pass from each

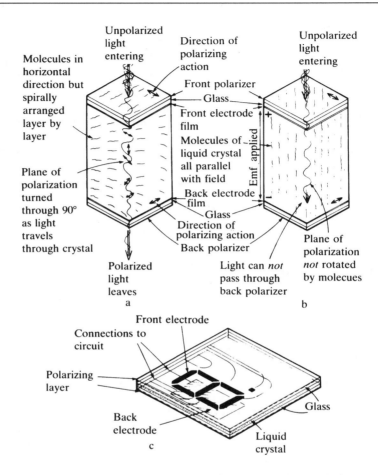

Figure 13.4 *Liquid crystal displays (a) When no field is applied, light passes freely through (b) When a field is applied, light is prevented from passing (c) A seven-segment LCD*

segment to the edge of the display and are there connected to a terminal strip. These tracks are too narrow to show up as black lines. When a p.d. is applied between any one or more of these segments and the back plane, the area appears black.

The illumination for an LCD comes from external sources such as daylight or a built-in filament lamp. The display may be viewed by reflected light, with a mirror behind it to increase contrast, or it may be viewed by transmitted light, with an illuminated panel behind it. The LCD display has the advantage that it can be viewed in full daylight, whereas an LED display is not bright enough. On the other hand, an LED display has the advantage in darkness. Since it produces its own light, an LED display requires an appreciable current, usually several tens of milliamps. By contrast, the current required by a LCD is only a few microamps. This is an advantage for displays on calculators and other portable battery-powered equipment, particularly for lap-top computers.

Cathode ray tubes

These are widely used as displays in oscilloscopes (p. 163), radar equipment (p. 262), TV sets and computer monitors. The way they work is described on p. 163.

14

From components to circuits

In the early days of electronics, a circuit was built by bolting or screwing the components to a chassis of sheet aluminium or wood and then joining their terminals by stout cotton-covered copper wire. A gradual reduction in the average size of components, plus the need for mass production has led to this technique being totally replaced by those described in this chapter.

Breadboard

A breadboard (sometimes called a **plugblock**) is used for building temporary circuits. It is useful to designers because it allows components to be removed and replaced easily. It is useful to the person who wants to build a circuit to demonstrate its action, then to reuse the components in another circuit.

A breadboard consists of plastic block holding a matrix of electrical sockets of a size suitable for gripping thin connecting wire, component wires or the pins of transistors and integrated circuits (ICs). The sockets are connected inside the board, usually in rows of five sockets. A row of five connected sockets is filled in at the top right of Figure 14.1. The rows are 2.54 mm apart and the sockets spaced 2.54 mm apart in the rows, which is the correct spacing for the pins of ICs and many other components. On some designs of board, longer rows of sockets occur along the edges of the board, usable for power supply rails. In Figure 14.1 the sockets on the extreme right, although spaced in groups of five are all connected together. The figure shows how a simple circuit is built up on the board. For example, the $+3$ V supply is connected to R1, to VR1 and (through a wire) to pin 7 of IC1. Pin 6 of IC1 is connected to the LED (D1), which is connected through R3 to the -3 V supply. Note the gap in the centre of the board, bridged by the IC. This ensures that opposite pins of the ICs are not connected.

Once a circuit is assembled on the breadboard it is tested. If its performance is found to be less than perfect, it is easy to substitute resistors or capacitors of different values. It is also easy to replace components suspected of being faulty. If sections of the circuit need to be isolated to investigate faulty operation, this may readily be done by removing one of the component terminal wires from its socket.

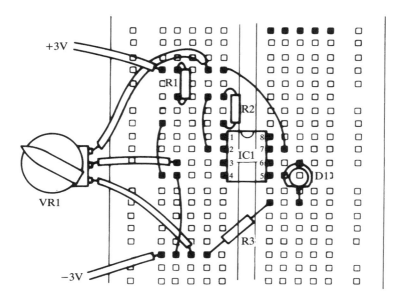

Figure 14.1 *A breadboarded circuit*

Breadboards are ideal for building and testing relatively simple circuits. Although they can, in theory, be used for complicated circuits, the board soon becomes covered with a nest of wires so that it becomes very difficult to trace the connections. If one of the wires is accidentally removed from its socket it is often difficult to find the correct socket in which to replace it.

Stripboard

Stripboard is one of the commonly-used types of prototyping board. These boards are intended for permanently assembling one-off circuits, especially prototypes. The board is made from insulating material, usually a resin-bonded plastic or fibreglass. One side has parallel copper strips on it, spaced 2.54 mm apart. There are holes bored in these strips, also 2.54 mm apart (Figure 14.2). Components are placed on the other side of the board with their wires bent to pass through the holes. The wires are soldered to the copper strips, the projecting ends being cut off to make the assembly neater.

The strips correspond to the rows of sockets on breadboards, allowing several components to be joined together, but a strip has many more than five holes so numerous connections are possible. Strips can be cut into shorter lengths where is it necessary to use one strip for making several different connections. A circuit built on stripboard has permanently soldered

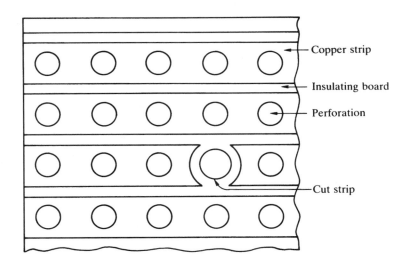

Figure 14.2   *A portion of stripboard, as seen from the strip side*

connections, but it is not difficult to remove a component and solder in one of a different value, or even to completely modify part of the circuit. A prototype circuit may be built and tested stage by stage.

There are various specialized designs of stripboard, some intended for use with integrated circuits. Others have a built-in connector pad at one end. Specially shaped and laid out prototype boards are available for assembling prototype cards for computers.

Printed circuit board

A printed circuit board (or PCB) begins as a board of resin-bonded plastic or fibreglass coated on one or both sides with a continuous layer of copper. It is then etched (see below) to produce a circuit layout, with pads to which components are soldered, and tracks to provide the required connections between the pads (Figure 14.3). Holes are drilled in the pads to accept component leads or pins and occasional wire links. During assembly all that is necessary is to insert the leads or pins into the holes, solder them to the pads and trim the surplus leads. There is no 'wiring up' of the board, other than connecting it to off-board components, as the wiring stage is taken care of when the layout is designed.

PCBs may be single-sided or double-sided, but the components are usually mounted on only one side, to make assembly easier. Connections between tracks on opposite surfaces of the board are made where necessary by drilling

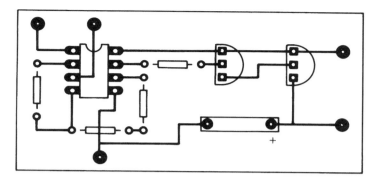

Figure 14.3 *Printed circuit board, showing square, rectangular and circular pads joined by tracks (the thick lines). The outlines of components, which are mounted on the opposite side of the board, are drawn in thin lines. The components comprise an 8-pin IC, an electrolytic capacitor, two transistors and four resistors. There are five terminal pins for off-board connections*

holes and soldering a double-headed rivet-like **via** to the matching pads on each surface.

The first stage in manufacturing a PCB is to design the layout. Special computer programs are available to assist with this, though a simple circuit can well be designed by hand. The aim is to arrange the components so the connections between them are as short as possible and that there are as few places as possible at which tracks need to cross each other. Naturally, tracks can not actually cross. Crossings can often be avoided by running tracks beneath components (under the IC and the resistor in Figure 14.3), and between adjacent pins on a component. Where a crossing is unavoidable, we use a wire link or, in the case of a double-sided board, route the track on to the other surface. When the design is complete it must be thoroughly checked because it is difficult to correct errors at the later stages.

The layout is transferred to the board by printing its design in etch-resistant ink. This has the big advantage that thousands of identical boards can be manufactured using the same printing plate. The board is then immersed in a bath of etchant solution. This removes the copper from all areas of the board except those covered by the resistant ink which is know as '**resist**'. At the end of etching, we are left with the pads connected by tracks. The resist is removed by a second bath. Usually the copper is tinned to protect it from corrosion and to make it easier to solder to. Finally, the board may have a lacquer layer printed on all over except for the pads, as corrosion protection.

Next the pads are drilled to accept the wire or pins of the components. The spacing of the holes is standardized so that components of standard size can be dropped by machine into the holes. Often special short-leaded resistors are used to make it easier to insert the wires automatically and to eliminate the need for clipping off the excess wire later. The components may be soldered in place by hand, but more often the board is passed through a bath of molten solder in which a 'wave' of solder sweeps across the board, soldering all junctions as it goes.

This is a technique specially suited to mass-production of items such as TV and radio sets and computers, and the description above outlines the production-line stages. The primary advantage of PCB manufacture, in addition to the high rate of circuit production, is that it does not require skilled operators to hand-wire the connections between components. Hand-wiring is always subject to a small percentage of mis-connections. If the original layout is correct, it can be guaranteed that every board will be correctly 'wired'.

Although the technique was primarily developed for mass-production, the PCB has found favour with the home constructor. Many electronics magazines sell PCBs of the projects which they have published, and most electronics kits include a PCB. Errors of wiring are eliminated, but it is still necessary to make sure that all the components are correctly inserted and that the solder connections are properly made. The home constructor can make one-off PCBs using one of several techniques to apply the resist. The simplest method is to draw the tracks by hand using a special lacquer pen. A neater method is to use rub-down pads and tracks, which are similar in use to rub-down lettering.

A popular method is to use copper-clad boards which are coated with a layer of photoresist. A positive transparency of the layout design is placed on the surface, which is exposed to UV light. Positive transparencies are obtained by hand-drawing on film, or using rub-down pads and tracks on film, or by photocopying on film from the printed designs published in books or magazine. The exposed board is then developed, rather like a photographic print, so that the photoresist is removed except where the tracks and pads are to be. After the resist pattern has been created on the board, the rest of the procedure is much like that in the industrial process, except that the home constructor is more likely to solder the connections individually and not use a solder bath.

To sum up, PCBs offer reliable mass production and freedom from wiring mistakes. The only disadvantage, which is not usually important in large-scale production of successful designs, is that it is difficult to alter or modify the circuit once it has been committed to a PCB.

Integrated circuits

On p. 81 we described how a transistor is fabricated on a slice of silicon. Resistors, capacitors and other components can be made in a similar way, as well as the connections between them so that a complete circuit can be produced on a single chip. This is known as an **integrated circuit** or IC.

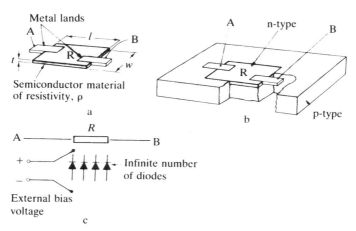

Figure 14.4 *Fabricating solid-state resistors (a) A resistive semiconductor layer with metal lands attached (b) A resistive layer diffused into a silicon chip (c) The equivalent circuit of the isolation technique*

Figure 14.4 shows how the resistors of an IC are made. The resistance of a volume of semiconductor depends on its length *l*, its thickness *t*, its width *w*, and on the resistivity ρ of the semiconducting material. Given these quantities we use this equation to calculate the resistance:

$$R = \frac{\rho l}{tw} \, \Omega$$

As stated on p. 65, the resistivity of semiconductors is high relative to that of metals. This means that even a microscopically small resistor made in this way can have a suitably high resistance. To make the resistor, a volume of semiconductor of specified dimensions is doped, using the masking techniques described earlier. The resistivity depends upon the amount of doping. Using another mask, areas of metal, known as **lands** are deposited on the chip to act as contacts. As we shall see later, the lands may be extended to make contact between the resistor and adjacent components.

The figure shows a resistor of n-type material formed in a slice of p-type material. Conduction is possible from the n-type to the p-type so the resistor is not electrically isolated from the substrate (the p-type). This could cause problems when several resistors and other components are fabricated close together on the same chip. The answer to this is to bias the substrate to be at a lower potential than the resistor. This, in effect makes the resistor-substrate junction a reverse-biased p-n junction, preventing the flow of current between resistor and substrate. Resistors can also be made of p-type material on an n-type substrate.

Crystals for ICs

ICs are produced on silicon wafers cut from a large silicon crystal. The diameter of a standard silicon crystal is 8 inches, but there is a trend to using larger crystals 12 inches or more in diameter.

To make a capacitor on an IC we rely on the principle of the varicap diode (p. 73). In Figure 14.5a a layer of p-type material is diffused into an n-type substrate and then a layer of n-type is diffused on top. Metal contacts are deposited at A and B. The diode is reverse-biased, creating a depletion layer (p. 69) which acts as the dielectric. When we use a varicap diode the idea is to vary the reverse bias to vary the width of the depletion layer and thus obtain a variable capacitor. In an IC, it is more usual to have a constant bias voltage and to produce capacitors of different fixed values, depending on the dimensions of the plates.

It is virtually impossible to fabricate an inductor in semiconductor material, but the functions of inductors can often be accomplished by sub-circuits built from an amplifier and a few capacitors and resistors (see p. 184).

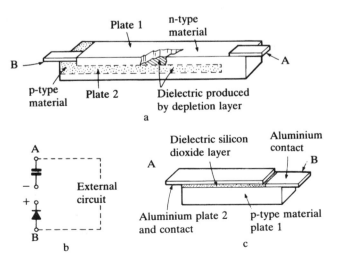

Figure 14.5 *Fabricating solid-state capacitors (a) A diffused capacitor which uses the depletion layer as the dielectric (b) The equivalent circuit (c) A capacitor which uses a silicon dioxide layer as the dielectric*

Integration

The way in which the basic components described in the previous section may be integrated into a single on-chip circuit is illustrated by the amplifier in Figure 14.6. The capacitor is formed by the aluminium contact A acting as one plate, a layer of silicon dioxide forming the dielectric and a layer of n-type material acting as the other plate. The aluminium contact B connects the capacitor to the base (the p-type region) of the transistor. One or two resistors (not shown, for simplicity) would be connected to contact B to bias the transistor into conduction. In the transistor the n-type layer nearer the surface is the emitter, and has a metal contact C for connection to the 0 V supply rail. From the lower n-type layer, the collector, there is a metal contact D from which the output signal of the amplifier is taken. This connects through a resistor to the positive supply at contact E.

IC manufacture

Integrated circuits may require several consecutive stages of masking etching and doping to build up the complex and interconnected semiconductor devices on the wafer. It becomes difficult to control the depth of doping accurately and to prevent dopant from spreading to areas where it should not be. In such cases an ion gun, accelerating ions of the dopant at high speed in a vacuum, is used to impregnate selected areas of the wafer with dopant. This is known as **ion beam doping**. Another technique which may be used at certain stages in production is growing a layer of n-type or p-type silicon on the surface of the wafer by condensing it from a vapour at high temperature. This is rather like the formation of crystals of hoar frost on a window or on the branches of a bush from air saturated with water vapour. This is known as **epitaxial growth**.

The final step in the manufacture of ICs is to test the individual circuits on the wafer, marking and discarding those which are defective. They are separated and each is mounted on a substrate. The entire amplifier of Figure 14.6 occupies an area of silicon less than 0.1 mm square, and many more complicated ICs are only 1 or 2 mm square. It would not be easy to connect these directly to an external circuit. The silicon chip is therefore mounted in a standard package, usually made of plastic but occasionally of ceramic materials or metal. Figure 14.7 shows some typical IC packages. The plastic **double-in-line** (DIL) package shown in Figure 14.7c is the most commonly-used type. Within

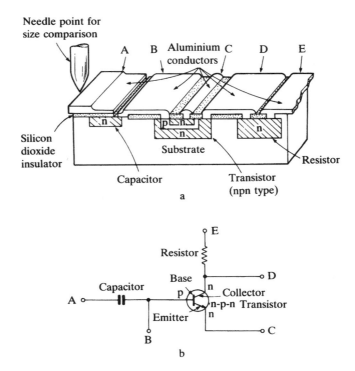

Figure 14.6   *A simple amplifier fabricated as an integrated circuit*

the package is a framework of conductors leading to the terminal pins. As shown in Figure 14.7d, the circuit chip is mounted in the centre. Terminal lands on the chip are connected to the terminal pins by thin gold wires soldered at either end. Special machines are used to perform the soldering quickly yet precisely. The package is then sealed.

Although DIL packages may vary in their number of terminals from 6 pins to 40 pins, the majority have 14 pins or 16 pins and look monotonously alike. Yet the variety of type numbers printed on the packages testifies to the wide range of types available. The different types of IC are outlined in the next section.

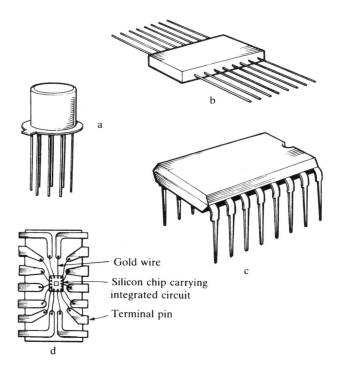

Gold wire

Silicon chip carrying
integrated circuit

Terminal pin

Figure 14.7 *Integrated circuit packages: (a) Metal can (b) Flat-pack (c) Double-in-line package (d) Double in-line package with the upper half removed to show the connections between pins and the circuit on the silicon chip*

Applications of integration

The main types of IC fall under the following broad headings:

Linear ICs

These include amplifiers of many kinds. Some are specially designed for audio applications and many of these operate at high power. Those of the highest power, often over 200 W, are provided with built-in heat sinks to dissipate excess heat. ICs for use in stereophonic audio systems have two identical amplifiers on the same chip, one for each channel. Radio-frequency amplifiers are also available in integrated form. One of the most versatile of amplifiers is the operational amplifier, or op amp. It is a high-gain amplifier with special features that make it useful in many applications, including audio amplification and instrumentation. The ways of using op amps are described in Chapter 17.

Logic ICs

The action of these is described in Chapter 19. They are also known as **digital ICs.** The early logic ICs provided only the simplest of functions such as the elementary logical operations NAND and NOR and required only a dozen or so integrated components on one chip. These ICs were followed by circuits of greater complexity such as counters, data storage registers, and arithmetic devices that could add two four-digit binary numbers. This level of complexity is known as **medium scale integration** (or MSI). As IC manufacturing techniques improved, and it became possible to fabricate the devices with smaller and smaller dimensions and so cram more and more on to a chip, MSI was followed by **large scale integration** (LSI) and later by **very large scale integration** (VLSI). Whole logical systems are fabricated on a single chip. A pocket calculator and a digital clock or watch have their complete logical circuit on a single LSI chip. The heart of the microcomputer is a single **microprocessor** chip in which all the logical and mathematical operations of the computer are performed. In the Pentium microprocessor chip there are almost 20 million p-n junctions, a good example of VLSI. Other examples are the **transputer** chips which comprise not only a microprocessor but also the control logic and memory, so that they are a complete computer on one chip. They can be manufactured so cheaply that it is economical to incorporate them in automatic equipment such as washing machines, industrial robots, microwave ovens, and automobile and aircraft control systems, even though one particular application does not make full use of the microprocessor's abilities. Most personal computer systems have a microprocessor to operate the video monitor and another to operate the printer as well as the main central processing unit (CPU) in the computer itself. The CPU has a specialized microprocessor assistant in the form of a **maths coprocessor** to assist it with complex mathematical operations and thus leave the main microprocessor free to cope with the more mundane tasks.

Special function ICs

It is impossible to describe the wide range of ICs that come under this heading. We have timer ICs that, with a few external components, can control the duration of time periods from a few microseconds to several days or even months. We have the phase-locked loop ICs that can be tuned to pick out and lock on to a signal of a particular frequency from a mass of other signals. There is the radio IC looking just like a transistor that, with a tuning capacitor plus two or three resistors, gives a complete radio set. There are the remote-control ICs that automatically generate a coded stream of pulses: when this code is received by the TV set, another IC decodes the message and controls changing of wavebands, the brightness and colour balance of the picture or the volume of the sound. There are special ICs for producing the many voices and rhythms of the electronic keyboard and synthesizer. There seems to be no limit to the possibilities of integrated circuits.

Gallium arsenide ICs

Although silicon is the most widely used base for integrated circuits, an alternative material is gallium arsenide. Devices based on this compound are much better for high-speed applications, including fast logic gates. They also use less power. Apart from the special safety precautions that must be taken because of the arsenic in the material, gallium arsenide has the further disadvantage that it is more difficult to work with, raising the cost of manufacture. In particular it is more difficult to produce an insulating oxide layer on gallium arsenide than it is on silicon. High temperatures are necessary, which may impair the structure of the crystal. A recent development is to grow a layer of nitride instead, a technique which is easier to carry out.

Advantages of integration

There are very few electronic items today that do not include at least one IC – or may even consist of just one IC. The development of integration techniques has brought about a major revolution in electronics and consequently in many aspects of our lives. The major advantages of integrated circuits are:

- Miniaturization – circuit boards and therefore electronic equipment can be smaller
- Low cost – the simpler ICs cost no more than a single transistor
- Production of circuits using ICs is more easily automated, leading to lower costs
- Use less power (especially CMOS); an advantage for portable equipment
- Greater reliability
- Longer life
- Easier servicing – instead of testing individual components, simply replace the IC
- Small size of transistors and the very small distances between components on the chip mean that the circuits are faster acting and can be operated at much higher frequencies
- Components that are close together on the same chip are all at the same temperature; this improves the stability and precision of certain types of circuit.

Surface mount technology

Although integration has much to offer, there are still many reasons for build-ing circuits or parts of circuits from discrete devices on PCBs. But, with the increasing sophistication and complexity of circuits on the one hand and the demand for compact, portable equipment on the other hand, any technique that reduces the physical size of components and therefore that of the PCB, is welcome. One such approach is **surface mount technology**, known as SMT for short. This is widely used nowadays in products that need to be small and portable such as pocket telephones and lap-top computers. It is also used for larger equipment in which size reduction gives benefits, for example in personal computers, to reduce the amount of desk space occupied.

Surface mount devices (or SMDs) do not have wire leads to pass through holes drilled in the PCB. Instead they have terminals which allow them to be soldered directly to the board on the same side as the component. Since no holes are needed, one tedious and time-consuming step in PCB production is completely eliminated, with valuable savings in production costs. The other key feature of SMDs is that they are smaller than the equivalent conventional components. Figure 14.8 shows an SMT resistor. This is the standard size that is most commonly used, but resistors are also made half this size, and higher power resistors are made slightly larger. Typical capacitors (up to 100 nF) have the same form and size as the resistors. There is a limit to the amount of size reduction possible with electrolytic capacitors, but low working-voltage types are only about 4 mm in diameter and are up to 20 mm long.

Most of the conventional semiconductor devices are also available as SMDs. Figure 14.9 shows a typical transistor package. There are no special problems in producing this as the actual transistor is small (p. 81) and only needs putting into a smaller package. The terminal pins are shaped to make even contact with

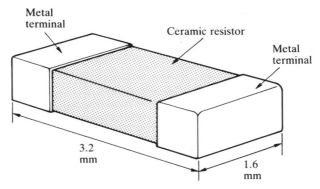

Figure 14.8   *An SMT resistor, type 1206*

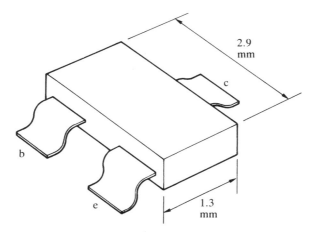

Figure 14.9   *An SMT transistor in the SOT23 package; diodes usually have the same package*

the surface of the board. Diodes and LEDs are produced in the same package, the LEDs being moulded in coloured transparent plastic. ICs too are basically very small so they can readily be fitted into a smaller package half the conventional size, with terminal pins like those in Figure 14.9, spaced only half the usual distance apart.

PCBs are prepared by masking and etching as for conventional through-hole construction (p. 134) with the important difference that no holes are drilled. The shapes and spacing of the pads are commensurate with SMD terminal sizes and spacing. Another important difference is that components may be mounted on both sides of the board, thus effectively doubling the number of components that may be accommodated on a board of given area. Any connections required between the two sides are made by carrying short leads around the edges. Before soldering, the board is printed with a pattern of solder paste so that there is a blob of paste on each pad. The paste consists of a mixture of minutely powdered solder and a soft resin. It is adhesive so that, when the components are placed on the board (either by hand or by machine) they remain in place. Blobs of adhesive may also be put on the board to hold the larger components such as electrolytic capacitors in place. In industrial production all components are soldered at once using a blast of hot air, or by passing the board through an oven at a controlled temperature. The solder melts and flows into the joints. On the small scale the solder is melted by touching the joint with a hot soldering iron.

SMT is more suited to the equipment manufacturer rather than to the home constructor, but several SMT projects have appeared in magazines. It is said that small is beautiful, and there is fun to be had in working with such miniscule components. The main requirements are a pair of forceps, a powerful magnifier, good eyesight, and a modicum of patience.

Computer simulations

Circuit design is something of an art. Although it is possible to predict the behaviour of a very simple circuit mathematically, there are so many factors to consider in a more complicated circuit that the calculations become impossibly convoluted. This is where the breadboard and subsequently the stripboard are so useful. Having decided on initial values for components, the final values may be arrived at by a process of trial and error.

Computer software is able to simulate the action of a circuit, using mathematical routines that would take far too long to perform manually or with an ordinary calculator. The intended circuit design is keyed in to the computer as a **netlist**. This lists the components, their values and other characteristics, and the connections to be made between the components. Some simulation programs have a facility by which, instead of keying in a netlist, the user draws the circuit diagram on screen, typing the values and other characteristics of the components on the diagram. The program uses this diagram to prepare a netlist automatically. The software is then asked to perform analyses of the behaviour of the circuit. These can be simple ones which, for example, produce a table of the potentials at all points in the circuit network; or they can be far more complicated, for example, showing the output waveform of an amplifier when a given signal is supplied to it. The action of the software is the same as having a real circuit on a breadboard and a range of instruments such as voltmeters, signal generators and oscilloscopes (p. 183) with which to test it. Changing component values or connections is only a matter of quickly editing the netlist or diagram; there is no need to unsolder a component and solder in a different one. And, of course, software 'components' are never faulty, and can never become damaged or burnt out.

Transistor and IC manufacturers supply libraries of 'components' in data form which match the behaviour of their real components, so that it is possible to specify exactly which type number is to be used in the circuit. It is also possible to 'sweep' component values, such as resistance, capacitance, and gain over their full tolerance range so that one can be certain that a circuit will work whatever the actual value of any component used in the actual circuit. Similarly the operating temperature can be swept to make certain that the circuit works equally well whatever the ambient temperature.

Many programs have been written for circuit simulation but most are variants and enhancements of a program known as SPICE. This is short for Simulation Program with Integrated Circuit Emphasis, pioneered and developed at the University of California, Berkeley. It was intended for designing

ICs, which by their nature are not amenable to breadboarding, but has subsequently been extended to use with discrete components.

Using simulation software saves the designer considerable time, it virtually eliminates the need for breadboarding. The design is perfected on the computer and is immediately ready to transfer to the PCB.

15

Oscillating circuits

The output voltage of an oscillating circuit alternates between two peak values at regular intervals of time. It may switch sharply between the two extreme values, in which case we call it a **square-wave oscillator**, but the output waveform may have other regular shapes such as sawtooth, triangular or sine waves. The time taken for one complete oscillation or cycle is known as the **period** of the oscillation. The rate of repetition of the waveform is known as the **frequency**. The two are related by the equation:

$$\text{frequency} = \frac{1}{\text{period}}$$

In most, but not all, of the oscillators described in this chapter the timing of one cycle, and hence the frequency of the oscillations is determined by the time taken to charge and discharge a capacitor through a resistor. The first example illustrates this point.

UJT oscillator

One of the simplest oscillator circuits is based on a single unijunction transistor or UJT (Figure 15.1). When the power is switched on, current begins to flow through resistor R, gradually charging capacitor C. The p.d. across C rises steadily and the potential at the emitter of TR_1 rises with it. At a certain point, known as the peak point, current begins to flow into the emitter, and through the transistor to base-2(b_2). This flow of current rapidly discharges C, so the p.d. across it falls sharply. It falls to a low level, when the flow of current into the emitter ceases. This completes one cycle. The next cycle begins as current continues to flow through R and C is gradually re-charged. At the output 1, the potential is that across the capacitor. It rises relatively slowly and falls sharply, giving a sawtooth waveform. The rate of rise, and therefore the length of the period depends on the values of R and C. Large values of R and C give a slow rate of charge, a long period and a low frequency.

Another waveform is produced at output 2. As the current is discharged from the capacitor at the end of each cycle, and emerges from base-2, it passes through the resistor to the ground rail. This causes a brief increase of p.d. across the resistor. The output is a series of short high pulses.

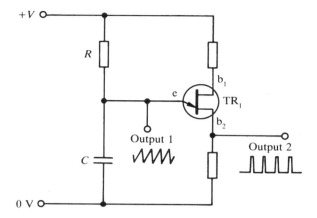

Figure 15.1 *A relaxation oscillator, based on a unijunction transistor*

In this oscillator the p.d. across the capacitor builds up slowly and is released almost instantly. An oscillator which has this characteristic is known as a **relaxation oscillator**. Several more examples of relaxation oscillators appear in this chapter.

Bistable circuit

This is not an oscillator but an explanation of how it works will help to explain the action of one of the oscillators described later. The word 'bistable' is applied to a circuit that is stable in either one of two states. There are several ways of building a bistable, one of which is shown in Figure 15.2. This is made from two transistor switches, cross-connected so that the output of one switch becomes the input of the other. The outputs and inputs are not connected directly but through resistors R_2 and R_3.

When the power is switched on, the circuit assumes one of its stable states. Assume that it begins with TR_1 switched on. If so, output A is at low potential and the base of TR_2 is at low potential. This means that TR_2 is off. Output B is therefore close to the supply voltage (call this 'high') and the base of TR_1 is receiving enough current to keep TR_1 switched on. The circuit is stable in this state and remains in this state indefinitely. Note that output A is low and output B is high.

Now suppose that we connect input A (the base of TR_1) to the 0 V rail or in some other way give it a low input. Immediately TR_1 is turned off. Output A becomes high. This turns TR_2 on and output B goes low. Even though we may connect input A to the 0 V line only for an instant, it is now permanently low and the circuit is stable in this state. Note that output A is high and output B is

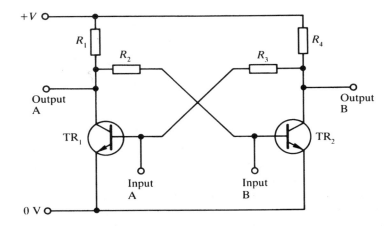

Figure 15.2   *A bistable circuit, or flip-flop*

low. In either state, one output is high and the other one is low. It is not possible for both to be high or both to be low.

Connecting input A briefly to the 0 V line has made output A high and output B low. Connecting input A to the 0 V line again has no effect, as the base of TR$_1$ is already at low potential. But see what happens if we connect the other input (B) briefly to 0 V. This turns off TR$_2$, output B goes high; TR$_1$ is turned on, output A goes low. The circuit has returned to its original state. As we connect the inputs A and B to 0 V alternately the circuit switches between one stable state and the other. It 'flips' and it 'flops', which is why this type of circuit is also called a **flip-flop**.

If we set up this circuit and leave it for someone to play with, we can always tell which input was most recently grounded. The circuit 'remembers' which input was most recently connected to 0 V. So although this circuit is not an oscillator it is a very useful one. Flip-flops are a way of storing data, and are used in the memory circuits of computers.

Monostable circuit

The term 'monostable' implies that this circuit is stable in only one state. Like the bistable it has two states but it is stable in only one of them. When it is put into the unstable state it sooner or later returns to its stable state. The circuit (Figure 15.3) is very similar to that of the bistable. It consists of two cross-connected transistor switches, but instead of linking them both through resistors, only one link is a resistor and the other link is a capacitor. This makes time a factor in the operation of the circuit.

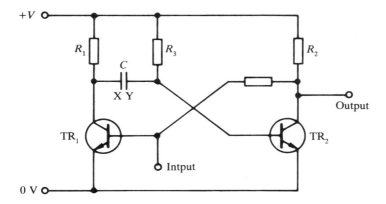

Figure 15.3 *A monostable circuit*

When the power is switched on, the circuit goes into its stable state with TR_1 off and TR_2 on. The collector of TR_2 (and also the output) is therefore low. No current flows to the base of TR_1, so TR_1 is off. Now consider the p.d. across the capacitor. Plate X is high, almost at the supply voltage. Plate Y is at about 0.6 V, the typical base-emitter voltage of a transistor. If the supply voltage is 6 V, the p.d. across the capacitor is about 5.4 V, with X positive of Y.

Now a brief positive pulse is applied to the input, for example, by touching the input to the $+V$ rail. This turns TR_1 on and the potential of plate X falls rapidly to (say) 1 V. This is a drop in potential of about 5 V. The capacitor acts to keep the p.d. across it constant. If X drops from 6 V to 1 V, the potential of Y drops by 5 V from 0.6 V to -4.4 V. This turns TR_2 off. The output goes high instantly.

With plate Y at -4.4 V, current flows through R_3, gradually charging the capacitor. The potential of plate Y (and also the potential at the base of TR_2) gradually rises. The rate of rise depends on the value of R_3 and of the capacitor. Eventually the potential of Y reaches 0.6 V and then current begins to flow into TR_2 turning it on. As soon as it starts to turn on, the output falls to low, and this turns TR_1 off. Its collector potential goes high, raising the potential of plate X almost to the supply level again. This forces the potential of plate Y up, completely turning on TR_2. While the capacitor was charging, the circuit was unstable, but now it is stable again.

The action of this circuit is that a brief high pulse to the input causes the output to go high for a period of time depending on the values of R_3 and C. In a practical monostable of this type, the output may be high for several seconds or even minutes. This circuit is not an oscillator but has many applications as a pulse stretcher (converting a short pulse into a much longer one) and as a delay circuit.

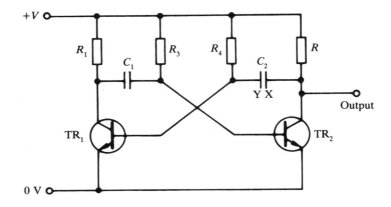

Figure 15.4 *An astable circuit*

Astable circuit

The word 'astable' means 'not stable'. This circuit is not stable in either state; it switches continually from one state to the other. It oscillates. The circuit (Figure 15.4) is very similar to that of the bistable and monostable, the difference being that the connections between the transistor switches are *both* made by way of capacitors. As in the other two circuits, at any instant one transistor is on and the other is off.

We begin with TR_1 off and TR_2 on. Output is low. If TR_2 has only just been turned on, the output has only just gone low and the potential of plate X of C_2 has fallen sharply. As in the previous description, this takes the potential of plate Y and that at the base of TR_1 to a negative value. TR_1 is firmly off, but it does not remain in this state. Current flows though R_4, gradually recharging plate Y. When the potential at Y and the base of TR_1 reaches 0.6 V, TR_1 begins to turn on. The potential at its collector starts to fall; this pulls down the potential on both plates of C_1, making the base of TR_2 negative and turning it off. Now the circuit is in its other state, but it does not remain in this state either. Current flows through R_3, gradually raising the potential at the base of TR_2 until it turns on again. The circuit returns to its original state.

This circuit automatically oscillates between its two unstable states at a rate determined by the values of R_3, R_4, C_1 and C_2. If the values are such that the circuit oscillates at a few hundred or thousand hertz, it can be used to generate audio-frequency signals. If it oscillates more slowly, it can be used for flashing warning lamps.

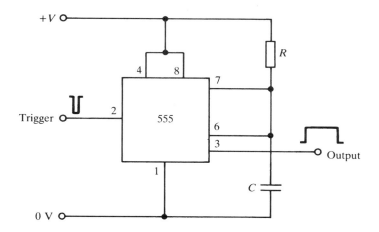

Figure 15.5 *A monostable circuit, using a 555 timer IC*

The 555 timer IC

This must be one of the most useful ICs ever invented. Figure 15.5 shows it being used as the basis of a monostable circuit. As can be seen, the circuit requires only three components instead of the seven components of Figure 15.3. This makes it much simpler to wire up. Also, because of its design, it gives more precise timing intervals than the monostable of Figure 15.3, and (in its CMOS version, the 7555) can be used to produce output pulses as short as a few microseconds up to as long as several hours. Once again, the timing relies on the charging of a capacitor through a resistor and the pulse length is set by choosing suitable values for R and C.

When power is switched on, current flows through R and begins to charge C. A circuit inside the IC, and connected to pin 6, detects the p.d. across the capacitor. When this has risen to one third that of the supply voltage ($V/3$), the current is then diverted into pin 7 and so to ground. The capacitor remains charged to $V/3$, and the output is 0 V. The timer is triggered into action by a brief low pulse applied to pin 2. Instantly, the output goes high. Current no longer flows into pin 7, but into the capacitor, increasing the charge on it until it reaches two-thirds of the supply voltage ($2V/3$). When pin 6 detects this level, the output goes low. At the same time pin 7 once more admits current, discharging the capacitor rapidly to $V/3$ again.

The output is high for as long as it takes to charge the capacitor from $V/3$ to $2V/3$. Since the rate of charging is proportional to V, the time taken to charge is independent of V. This is an important feature of this timer. The supply voltage (within the range $2V$ to $18V$ for the CMOS version) makes no difference to the length of the output pulse, and it is not affected by the battery being flat.

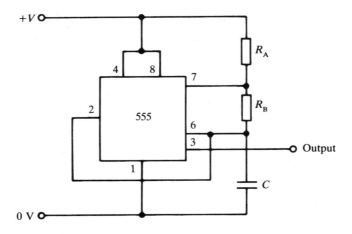

Figure 15.6 *An astable circuit, using a 555 timer IC*

The 555 timer can also be used in an astable circuit (Figure 15.6). This requires two resistors. The capacitor charges through both resistors but, when it has charged to $2V/3$ it is discharged through R_B only. The trigger pin is connected to pin 6 so the IC is re-triggered as the charge falls to $V/3$. The timer is triggered repeatedly, giving a square output waveform. The frequency is determined by selecting suitable values for R_A and R_B. Since it charges through R_A and R_B but discharges only through R_B, the length of the 'high' part of the cycle is longer than that of the 'low' part. One version of this IC can produce frequencies up to 2 MHz, and all versions can run at very low frequencies with cycles lasting several hours. The frequency is unaffected by the supply voltage.

Before leaving this topic we should mention another IC, the LM3909, that is used for building astable circuits (Figure 15.7). Here it is being used as an LED flasher, though it has other applications at frequencies around 1 Hz.

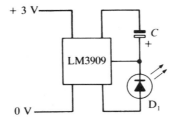

Figure 15.7 *An astable circuit, using a special IC*

Crystal oscillator

Here is an oscillator which depends not on the timing of a resistor-capacitor combination but on the mechanical properties of a quartz crystal. Crystals can be cut so as to vibrate at a precisely fixed natural frequency. Electrodes are deposited on two opposite faces. The crystal is made to oscillate by connecting it into a resistor-capacitor circuit with approximately the same frequency as the crystal. Owing to the piezo-electric effect (p. 107) the oscillations of the circuit make the crystal vibrate at its natural frequency. The p.d.s developed across the crystal force the circuit to oscillate at the same frequency.

Crystals are used for high-precision oscillators, such as those used for driving clocks and watches. The crystal used in a watch may have a natural frequency of 32.768 kHz (note the five-figure precision). This is a high frequency, which is inevitable since a small object such as a crystal can not have a low natural frequency. The watch has circuits which divide this frequency. At each stage the frequency is halved so that with 15 stages of frequency division the frequency is divided by 2^{15}. This equals 32 768, so the resulting frequency is 1 Hz. One cycle per second is ideal for driving clocks and watches. The precision is such that a fairly inexpensive crystal-controlled clock or watch is expected to be accurate to within 15 seconds per month. Crystal oscillators are also used to produce the high-frequency oscillations of radio transmitters.

Sine-wave oscillators

The principle of a resonant circuit was described on p. 62. If such a circuit is fed with oscillations of the correct frequency it resonates strongly, and the waveforms found in the circuit are all sine waves. Such a circuit is the basis of a sine-wave oscillator. Here we describe just one of several different designs of oscillator, the **Colpitts oscillator** (Figure 15.8). The resonant circuit consists of an inductor L_1 and two capacitors VC_1 and C_1. VC_1 is a variable capacitor which can be adjusted to tune the circuit to resonate at any required frequency. The resonant circuit is placed in the collector circuit of TR_1. Oscillations in the collector current (produced in a way we shall describe shortly) cause the circuit to oscillate at the resonant frequency.

To see how the transistor is made to produce oscillations, we must look at the connections to its other two terminals. The base of the transistor receives its current from a potential-divider, consisting of two resistors R_1 and R_2. The base potential is held very steady by the action of the large-value capacitor C_2. We can consider the base potential as being constant. The emitter of TR_1 is connected to a wire running from the junction of the two capacitors in the resonant circuit. In effect, we are tapping off a small signal from the resonant circuit and feeding it to the emitter. We call this **feedback**. The effect of this feedback is to cause the emitter potential to oscillate at the resonant frequency.

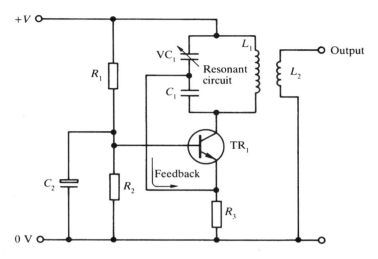

Figure 15.8   *A Colpitt's oscillator circuit*

With the base potential fixed and the emitter potential oscillating, the base-emitter potential difference is oscillating. The oscillations are only a few millivolts but are amplified by the transistor action, causing a larger oscillation in the collector current. This keeps the circuit resonating. In short, we tap part of the signal from the resonant circuit, amplify it, and use it to maintain the circuit in oscillation. The action is self-sustaining and the circuit oscillates continuously.

The output from the circuit is obtained from a second coil L_2 wound on the same core as L_1. In practice L_1 and L_2 make up a transformer. Oscillations in the magnetic field of L_1 induce oscillating currents in L_2, at the same frequency.

16

Test equipment

The moving-coil meter

Nowadays, most meters have a digital readout but the old-fashioned moving coil meter with its needle moving over a graduated scale still has many uses and is preferred by some people. Its operation relies on the fact that if a coil carrying a current is placed in a magnetic field, a force will act upon the coil. In Figure 16.1 a coil L is wound on a former of aluminium over a cylindrical iron core. The former is mounted on pivots so that it can rotate between the poles of a permanent magnet M. There is only a narrow air gap between the poles and the former, so the magnetic field in which the coil turns is radial in direction and uniform in strength for any position of the coil. The pointer P is carried by the coil. The turning motion of the coil is opposed by a phosphor bronze spring S.

When a current passes through the coil, the coil tends to turn on its axis against the opposing force of the spring until it takes up an equilibrium position which is proportional to the current flowing through the coil. As the coil turns, the pointer attached to it moves across a graduated scale. The amount of

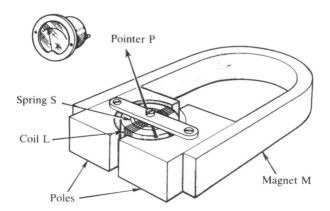

Figure 16.1 *A moving-coil meter movement*

157

turn is proportional to the current passing through the coil so the scale is graduated by comparison with a standard current meter to show the current, usually in microamps or milliamps.

Measurement of current

The elementary moving-coil ammeter that we have just described may have its current range extended by using a shunt. This is a low-value resistor wired in parallel with the coil of the meter. The way it works is shown in Figure 16.2a. Suppose for example, that we have a meter in which the needle is deflected the whole way across the scale when a current of 1 mA is flowing through the soil. We say that the **full scale deflection** (or f.s.d.) of the meter is 1 mA. In the figure, such a meter is shown with a shunt across it. The resistance of the shunt is 1/1000 of the resistance of the coil. When a current reaches the junction between the meter and the shunt, 1/1000 of the current passes through the coil and 999/1000 of it passes through the shunt. If the needle is fully deflected, we know that 1 mA is passing through the coil. But this is only 1/1000 of the current in the circuit. So the current in the circuit must be 1000 times greater, which is 1 A. The shunt has converted the 1 mA meter to a 1 A meter.

Note that a current meter (an **ammeter**) has very low resistance, equal to that of the coil and shunt in parallel. When we are measuring current we connect the meter, and its shunt if there is one, so that all the current flowing in the circuit must pass either through the coil or through the shunt.

Measurement of p.d.

In Figure 16.2b we are trying to measure the p.d. across a resistor which has a current passing through it. Given a coil with resistance R and given that the meter (in this example) has an f.s.d. of 50 µA, we can calculate from Ohm's Law (p. 20) that, when the needle is at full-scale, the p.d. across the meter is:

$$V = 50 \times 10^{-6} \times R$$

If the coil resistance is 200 Ω, the f.s.d. is $V = 50 \times 10^{-6} \times 200 = 0.01$ volts. We have a meter capable of measuring p.d.s up to 0.01 volts. This is a very limited range but we can easily extend it by putting a high-value resistor in series with the meter coil, as in Figure 16.2b. Contrast this with a shunt which is a low-value resistor in parallel with the coil. Suppose that the value of the series resistor is $R = 99$ times that of the coil. Given that the same current passes through the coil and through the resistor, the fall of p.d. across the resistor is 99 times that across the coil. The fall in p.d. across the coil and resistor together is 1000 times that across the coil alone. At f.s.d., when the p.d. across the coil is 0.01 volts, the p.d. across coil and series resistor is $1000 \times 0.01 = 10$ volts. We

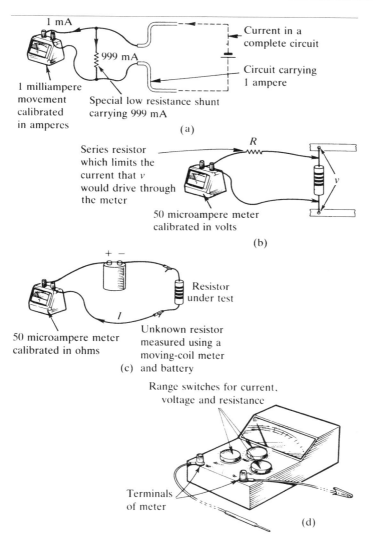

Figure 16.2 *Testing with meters (a) Measuring current with a milliammeter (b) Measuring p.d. or voltage with a microammeter converted to operate as a voltmeter (c) Measuring resistance with a microammeter (d) A typical moving-coil multimeter*

have increased the range of the meter to a f.s.d. of 10 V. By using series resistors of other values we can obtain any f.s.d. we require.

Note that a p.d. meter (a **voltmeter**) has very high resistance, equal to that of the coil and series resistor in series. When we are measuring p.d. we connect the meter, and its series resistor if there is one, to the two relevant points in the circuit. Only a very small current passes through the meter.

The descriptions of current and p.d. measurement apply to direct current and d.c. voltages. To convert d.c. into a.c., we wire a diode in series with the meter. Usually the frequency of the a.c. is too high for the needle to be able to follow the rapid changes in current or p.d. It shows an 'average' reading. This is less than the peak value (amplitude) and it can be shown that it is actually 0.707 times the peak. This is often known as the **root mean square** (or r.m.s.) value. Meters intended for measuring a.c. may have the scale calibrated so as to read peak and r.m.s. values.

Measurement of resistance

Figure 16.2c shows the principle of using a moving-coil meter to measure resistance. A cell provides a known p.d., V. The meter measures the current, I. From these two values we may calculate the resistance R, using the Ohm's Law equation:

$$R = \frac{V}{I}$$

Rather than use a cell directly as in the figure, it is better in practice if we use a zener diode or bandgap reference to provide a stabilized p.d. (p. 89). If the value of V is fixed in this way, the equation shows that R is proportional to $1/I$. We can calibrate the scale of the meter directly in ohms. Note that the scale reads from 'infinite' resistance at the zero end (very large resistance means very low current) to some minimum value of resistance at f.s.d. The ohms scale is reversed and is not linear, making accurate reading difficult.

There is a source of error in this circuit, for we are ignoring the resistance of the meter itself. The value R is really the sum of the resistance of the meter and that of the resistor under test. Ammeters usually have a relatively low resistance so the meter resistance can usually be ignored, but there will be serious errors if the resistor under test has a low value.

Multimeters

Moving coil meters used in scientific laboratories are often made with a single shunt or series resistor for measurements over a single range. In electronics, we more often use a meter which has switchable ranges. This is known as a **multimeter**. The multimeter has two probes wires, positive and negative, attached to its terminals (Figure 16.2d). These may end in a crocodile clip as shown on the right or in a test prod as shown on the left. Usually there are one or more rotary switches used to bring various shunts or series resistors into the measuring circuit. It may also have a separate input terminal for a.c. measurements with a diode leading to the measuring circuit, or there may be a d.c./a.c. switch to bring the diode into action when needed.

Digital meters

Instead of a dial with a pointer, a digital meter has a display consisting of four, possibly more, digits, together with polarity symbols (+ and −) and a decimal point that automatically appears between the correct pair of digits. The display is nearly always a liquid crystal display and is updated at intervals of a few seconds. Although the display may be the most obvious difference between this type of meter and the moving-coil meter, there is a much more fundamental difference between the two types. The input terminals of the moving-coil meter connect the test circuit to the coil and possibly a shunt or series resistor. The input terminals of a digital meter connect the test circuit to an operational amplifier (p. 175), usually one with FET inputs. The result of this is that the digital meter draws virtually no current from the test circuit.

The actual resistance of a moving-coil meter depends to a large extent on the quality of the meter movement. A low-cost meter usually has a coil of relatively low resistance and the series resistor has a correspondingly low value. Such a meter draws an appreciable current from the test circuit. In Figure 6.3b, we showed how connecting a low-resistance circuit to a potential divider results in

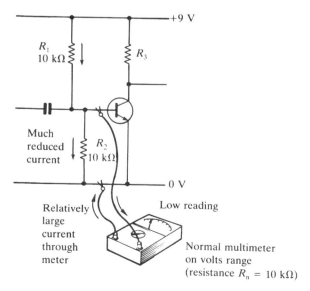

Figure 16.3 *A low-cost multimeter may give misleading results in some instances*

a fall in potential at the connection point. The same happens when a low-cost meter is used for measuring potential in a test circuit (Figure 16.3). The meter draws excessive current from the circuit, leading to a fall in potential at the point to which it is connected. It reduces the potential that it is trying to measure, giving a falsely low reading. This effect is much less important when using an expensive meter with a high-resistance coil, but with any moving-coil meter this effect can lead to errors. By contrast the digital multimeter with its FET input has almost no effect on the test circuit and an accurate reading of potential is obtained (Figure 16.4).

Digital meters are the product of the latest developments in electronic technology so it is to be expected that they will incorporate many features that are not available on the typical moving-coil meter. As well as a wide selection of voltage, current and resistance ranges (including accurate low-resistance range) many of these meters also provide for measurements of capacitance and frequency. A continuity tester, which produces a 'beep' when there is an electrical connection between the probes is almost a standard feature. The more expensive digital meters include circuits for testing diodes and transistors, including measurement of transistor gain.

In spite of the advantages of the digital meter, some engineers prefer the moving-coil meter for investigating fluctuating p.d.s and currents. Useful information may be gained by watching the way the needle moves across the scale. It is far from easy to extract this information from a set of rapidly changing digits. However, the more expensive digital meters have a way of presenting

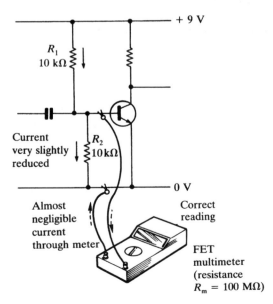

Figure 16.4   *A multimeter with FET input is required for accurate voltage measurements*

this information on their displays as a bargraph. Watching the right-hand end of the bar moving to and fro is equivalent to watching the needle of the traditional moving-coil meter.

The oscilloscope

One of the most useful test instruments in electronics is the oscilloscope, the heart of which is the cathode-ray tube. A typical tube is shown in Figure 16.5. The cathode ray tube consists of a specially shaped glass vacuum tube containing a number of electrodes. The first of these is the cylindrical cathode which is heated by a filament powered from a low-voltage supply. When the cathode is hot it emits electrons, which form an electron cloud around the cathode. But when the extra-high-tension power supply is switched on the electrons are accelerated toward the other electrodes which are held positive with respect to the cathode. These electrodes are the grid, a focusing anode and a main anode. The electrodes are cylindrical so the beam passes through them and strikes the tube face. This is coated with a phosphor which glows when hit by electrons, the number of electrons that hit the tube face (or screen) in a given interval determines the brightness of the glow. It can be seen in Figure 16.5 that the different potentials of the electrodes are obtained by tapping various points on a resistor. The main anode is positive with respect to the cathode, to provide the basic accelerating force. The grid is slightly negative of the cathode, removing some of the electrons from the beam. This acts to control the number of electrons reaching the screen and hence the brightness of the spot. The focusing

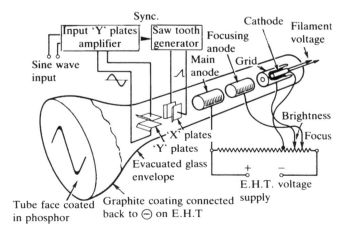

Figure 16.5   *The main parts of a cathode-ray oscilloscope*

anode is also negative with respect to the cathode. The focusing electrode is shaped so that it creates an electric field, which focuses the beam to a fine spot. This potential too can be adjusted to focus the beam precisely.

After passing through the main anode the electrons are travelling at very high velocity. After impinging on the screen they eventually find their way back to the cathode by way of a graphite coating on the inner surface of the tube. As they strike the screen they give up energy to the phosphor to make it glow brightly. If the electron beam is stationary the tube will have a stationary bright spot in the centre of the screen. The colour of the spot may be green, blue, orange, or some other colour, and it may have long or short persistence, depending on the composition of the phosphor. Different phosphors are used depending upon the purpose of the tube. The term 'phosphor' has nothing to do with the element phosphorus. It is used to describe any substance used for coating the screen simply because it glows or phosphoresces when activated by the electron beam. The term persistence is self explanatory. A long-persistence phosphor may glow for several seconds or more after bombardment. A short-persistence phosphor ceases to glow almost as soon as the beam is cut off. Most tubes used in oscilloscopes have a short-persistence phosphor, so that the rapidly moving beam draws a clear-cut picture on the screen. By contrast a tube used in a radar PPI display (p. 262) needs a long-persistence phosphor so that the images of objects located by the radar remain glowing on the screen until the next revolution of the antenna.

X and Y plates

Between the main anode and the screen the electron beam passes between two sets of deflector plates. The pair known as the X plates deflect the beam to left or right (in the x-direction on a conventionally plotted x–y graph). A time-base circuit produces a sawtooth p.d. which is applied between the plates to repeatedly deflect the beam steadily across the screen from left to right and then rapidly flick it back to the left. The effect of this is to produce a horizontal line across the centre of the screen. Given the right amount of persistence and a suitable repetition rate, the line appears to be constant.

The Y plates are situated above and below the beam and deflect it in a vertical direction. The alternating p.d. that is to be observed on the screen is fed to an amplifier, the Y-amplifier, which produces a corresponding alternating p.d. between the Y plates. The action of the sawtooth generator is synchronized with the input signal so that it repeats for each cycle of the incoming wave. Given the correct adjustments of the circuits, as the wave goes through one cycle the spot of light moves from left to right across the screen. Instead of drawing a straight horizontal line, the beam draws a graph of the waveform. In Figure 16.5 we show a sine-wave input and the resulting sine wave displayed on the screen (except that in reality the display would be a bright line on a dark ground). Given other waveforms such as square waves, triangular waves or the complex waveforms of audio signals the corresponding shape is displayed on the screen.

The screen of an oscilloscope is usually marked with a one-centimetre square grid to allow input voltages to be measured on the screen. The scale of the display can be set in both directions. The rate at which the beam scans horizontally across the screen can be set to a range of values so that 1 cm in the horizontal direction corresponds to 1 s, 100 ms, 10 ms, down to 0.1 µs. The shorter times allow signals in the radio-frequency bands to be displayed. The gain of the Y-amplifier can also be set so that 1 cm in the vertical direction corresponds to 20 V, 100 mV, 10 mV, down to 1 µV.

The above describes the simplest form of oscilloscope but other more elaborate models are available. Oscilloscopes often have a second beam with its own timebase and amplifier so that two signals may be observed at the same time and compared. There are also triple-beam instruments. Another refinement is trace storage, in which a single cycle of a waveform (which might be a single pulse of complicated shape) is captured and stored, then displayed on the screen to allow its features to be examined.

With the increased popularity of personal computers another approach has been made to observing waveforms. Much of the function of the cathode ray tube (hardware) can be taken over by a computer program (software). The signal to be examined is fed to an amplifier which is interfaced to the computer. The computer samples the signal at regular intervals, perhaps several million times per second and converts each voltage sample into its corresponding binary form. This data is then processed in the computer in a variety of ways. It can generate on its screen a display identical with that of the cathode ray tube. Also there are many other ways in which the data can be stored and analysed by the computer to obtain other information from it.

Signal and pulse generators

The simplest sort of signal generator is an oscillator, usually based on just one IC enclosed in a matchbox-sized case with a metal probe at one end. The frequency is about 1 kHz and its function is to inject a signal into a test circuit. Using an oscilloscope, a headset or the audio output stage (if it has one) of the circuit we can see if the signal is passing along from stage to stage in the circuit. If we find that it reaches a certain point but is not detected beyond that point, we know where to begin to look for faults.

Signal generators are also used for investigating the behaviour of circuits over a range of audio or radio frequencies. The amplitude and the frequency of the signal can be switched over several ranges. Frequency can be set from a few cycles per second up to several hundred megahertz. Usually a selection of waveforms is available including sine, square, triangular and sawtooth waves. A pulse generator is able to produce nominally square pulses of known duration and with specified rise times and fall times. The interval between pulses may also be adjusted.

17
Amplifier circuits

In this chapter we describe how transistors together with a few other components are used to build simple amplifier circuits. The amplifying action of a bipolar junction transistor was described on p. 82, but now we look at this more closely.

Transistor characteristics

There are various test circuits that are used to measure the way in which a transistor behaves. The results of one of the more useful of such tests is illustrated in Figure 17.1. The transistor is set up in the common-emitter connection as in Figure 9.11 (p. 94) with additions to allow the base current and the collector-emitter p.d. to be controlled and meters to measure the base and collector currents. First look at the lowest curve in the figure. The base current I_B is set at 50 µA and the collector-emitter p.d. is gradually increased from 0 V to 30 V. At each stage we measure the collector current I_C. Collector current begins at zero but rapidly increases to 10 mA. From then on, further increase in the collector-emitter p.d. produces virtually no further increase in I_C. The line is almost horizontal. If we repeat the test but make I_B equal to 100 µA (double the previous I_B) the shape of the curve is as before but now levels out with I_C equal to 20 mA (double the previous I_C). The same applies to the other two tests illustrated in Figure 17.1 In each test I_C levels out at a value that is 200 times that of I_B. Collector current is proportional to base current, with a **current gain** of 200.

The figure clearly demonstrates the amplifying action of the transistor. It also shows that, once the collector-emitter p.d. is greater than about 2 V the amplifying action is independent of the actual collector-emitter p.d. Increases in I_B result in proportionate increases in I_C but there is a limit to this, not shown in the figure. Once I_B has reached a given value, there is no corresponding increase in I_C. We say that the transistor is **saturated**.

For a transistor to amplify as shown in Figure 17.1, it must be in its **operating region**. The lower limit of this region is when the collector-emitter p.d. is very low and the transistor is operating on the steeply sloping part of the curve. The upper limit of this region is when I_B is sufficient to saturate the transistor.

167

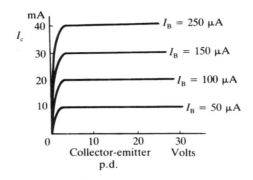

Figure 17.1   *The characteristic of a silicon junction transistor with a gain of 200*

Common-emitter amplifier

Figure 17.2 shows a single-transistor amplifier, using the transistor in the common-emitter connection. The power supply voltage must be large enough to produce a suitable collector-emitter p.d. and thus provide one of the conditions mentioned above for putting the transistor into its operating region. In a typical radio receiver or tape player the supply voltage is between 3 V and 9 V, although some receivers operate on a supply of only 1.5 V.

The next consideration is the **quiescent state** of the amplifier. This is the state of the amplifier when it is operating but when it is not receiving any signal to amplify. In the quiescent state there is a constant base current and a constant (and larger) collector current. The collector is at a constant output voltage, measured with reference to the 0 V line. It is usual for the quiescent collector voltage to be half the supply voltage. When a signal is being amplified the collector voltage rises and falls. If it is normally half-way between 0 V and the supply, it is able to rise and fall freely and equally in both directions. This

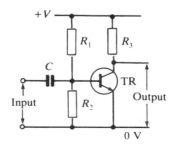

Figure 17.2   *A common-emitter amplifier*

makes it possible to have a large output signal without distortion. Usually, except in power amplifiers, the quiescent collector current is a few milliamps. Given this current, the value R_3 is selected to drop the voltage at the collector to the half-way level.

The transistor is **biased** into its operating region by the two **biasing resistors** R_1 and R_2. These act as a potential divider which provides sufficient base current to produce the required quiescent collector current. It is possible to bias a transistor with a single resistor either connected as R_1, or between the collector and base, but biasing with two resistors gives the amplifier greater stability.

The signal is fed into the amplifier through the **coupling capacitor** C (p. 43). An alternating signal arrives at the input and passes across the capacitor. It causes a small and varying current to be added to or subtracted from the constant base current supplied by R_1/R_2. In some circumstances it is possible to omit the capacitor and connect the amplifier to the signal source, but capacitor coupling is more often employed. For example, as we shall see in the two-stage amplifier, the quiescent output voltage (half the supply voltage) of the first stage would bias the transistor of the second stage into saturation if it were directly connected. The coupling transistor allows there to be a constant p.d. across the transistor without affecting the transmission of alternating signals across it.

Input impedance

The potential at the junction of the two resistors of a potential divider depends on their ratio. For example, it might be found that a suitable base current is provided by making R_1 equal to $220\,k\Omega$ and R_2 equal to $18\,k\Omega$. If the supply voltage is $9\,V$, the potential at the base is $9 \times 18000/238000 = 0.68\,V$. But the same potential could also be obtained if the resistors were $22\,k\Omega$ and $1.8\,k\Omega$, or even as small as $220\,\Omega$ and $18\,\Omega$. Using smaller resistors would not affect the biasing of the transistor but it would have a serious effect on the input signal. Low-value resistors would short-circuit most of the tiny signal current to the $0\,V$ or $+V$ rails. Little of it would pass to the base. Most of the signal would be lost.

If the input resistance of the amplifier is too low, the signal is mostly lost. The same idea applies when we consider the fact that the capacitor, together with R_1 and R_2, make up a high-pass filter (p. 62). With certain values of the capacitor and resistors, the low-frequency portion of the signal is blocked and never reaches the transistor. The combined resistive and capacitative effects on the input side are the **input impedance**. For the maximum signal to reach the transistor and be amplified, the input impedance must be as high as is feasible. This is achieved by making R_1 and R_2 as high as possible (if they are too high, the base current will not be large enough) and to select a value for the capacitor so that all required low frequencies are passed.

Output impedance

This is the impedance offered to the flow of current from the output side to the next stage of amplification or to a loudspeaker. In general, this should be as low as possible. The value of R_3 should not be high, otherwise it may be impossible for enough current to flow from the supply line through R_3 and on to the next stage. Similarly, if there is a coupling capacitor on the output side, the high pass filter that it makes with R_3 must pass all required low frequencies.

Input and output impedances vary with frequency. When specifying the performance of an amplifier, they are usually quoted for a signal frequency of 1 kHz.

Impedance matching

As explained above, if a circuit has low input impedance, a signal fed to it from another circuit is likely to be partially or almost wholly lost. Similarly a circuit with high output impedance will not produce sufficient current to drive a subsequent stage of low input impedance. It can be shown that for maximum transfer of power between two stages of a circuit, the output impedance of the first stage must equal the input impedance of the second stage. Impedance matching is essential if power is not to be lost. One way of achieving this is to select suitable values of resistors and other components at the output and input. Most instances in which impedance matching is required involve feeding a low-impedance input from a high-impedance output. This can be done by linking the two circuits by an emitter-follower amplifier (Figure 9.13).

Two-stage amplifier

Figure 17.3 shows an amplifier that has two stages, connected one after the other, **in cascade**. It does not simply amplify a signal and then amplify it still further. The function of the first stage, based on TR_1, is to amplify the voltage of the signal produced by the microphone. The function of the second stage, based on TR_2 is to amplify the current, making it sufficient to drive the loudspeaker.

The input from the crystal microphone is connected directly to the base. A crystal microphone has a high resistance so connecting it in parallel with R_2 does not have any appreciable effect on the potential-divider action of R_1 and R_2. It does not upset the biasing of TR_1 so there is no need for a coupling capacitor. TR_1 is a high-gain transistor to amplify the input voltage from the microphone.

The emitter of TR_1 is connected to the 0 V rail through a resistor VR. In the quiescent state, the current through VR generates a constant p.d. across VR,

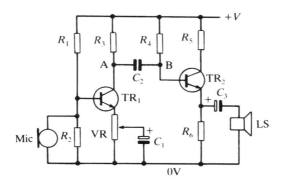

Figure 17.3 *A two-transistor audio amplifier*

raising the emitter potential by a few volts, perhaps to about 2 V. This means that R_1 and R_2 must bias the base to about 0.6 V more than this, in other words to about 2.6 V. R_3 is chosen to bring the collector potential (at A) half-way between the supply voltage and the emitter potential. This stage does not have to deliver a large current, so it need not have a high output impedance. R_3 can be a fairly high-value resistor. We discuss the function of C_1 later.

With the collector of TR_1 at a substantial potential, it is obvious that a capacitor C_2 must be used to couple this stage to the next. The next stage has only a single bias resistor R_4. R_5 and R_6 are chosen to bring the quiescent potential at the emitter to the half-way point. Their values are relatively low to enable them to pass relatively high currents. TR_2 is a low-gain transistor able to pass a substantial current without over-heating. It may require a heat-sink (p. 65). The low values of R_5 and R_6 give the second stage a low output impedance, which means it can deliver a large current to the final stage, the loudspeaker. Loudspeakers typically have very low resistance, sometimes as low as 4Ω and rarely higher than 100Ω so they need a large current to drive them. As the current through R_6 fluctuates, a varying p.d. is developed across it. This produces a signal at the junction of R_6 and the emitter and hence at C_3. The signal passes across C_3 and drives the loudspeaker.

The action of C_1 is as follows. When the potential at the base of TR_1 rises, base current increases and so does the collector current. The increased collector current passes through TR_1 and VR. Increased current through VR generates an increased p.d. between its ends. Its 'lower' end is fixed at 0 V, so the potential at its 'upper' end rises. The factor that decides how much base current flows to TR_1 is the base-emitter p.d. We have just explained that a rise in base potential results in a rise in emitter potential. As the base current rises and the base-emitter p.d. is increased, the emitter potential rises too and the base-emitter p.d. is reduced. The actions are in opposite directions. We call this action **negative feedback**. It could happen that the rises in base potential and emitter potential were equal. In such an event the base-emitter p.d. remains constant, so the base current remains constant and likewise the collector current. The signal is completely damped out. No sound would be heard.

The function of C_1 is to absorb part of the change, so that the amount of feedback can be controlled. The capacitor taps off part of VR. Fluctuations in the p.d. across the tapped-off part are damped out by the action of the capacitor. Only the fluctuations in the part of VR between the emitter and the tapping are available as feedback. By allowing a certain amount of feedback the gain of the amplifier is reduced. This seems to be a drawback but it is compensated for by the fact that the fidelity of the amplifier is much improved. Its gain is constant and is not affected by the frequency or amplitude of the signal or by temperature as it is when there is no feedback.

FET amplifier

As explained on p. 76, an FET has high input impedance. This makes it very suitable for use as the first stage on an amplifier that is to receive its input from a very high impedance source, such as a capacitor microphone (p. 108). In the circuit of Figure 17.4 the current through R_2 causes a p.d. to develop across R_2 when it is in the quiescent state. For reasons given earlier, the p.d. is usually half the supply voltage. In other words the source of TR_1 is at a positive potential. The gate is held at zero potential because of R_1. Consequently the gate potential is negative with respect to the source potential, as required for the operation of the transistor (p. 75).

The signal passes through C_1 to the gate, making the gate potential fluctuate and resulting in a varying current through the transistor. Note that although the gate of TR_1 offers high input impedance, the impedance of the amplifier is that of R_1, typically $1\,M\Omega$, which is high enough for most applications. The transistor and R_2 both have low resistance (a few hundred ohms) so the output of this amplifier has low impedance. Capacitor C_2 couples this amplifier to later stages which may be based either on bipolar or field effect transistors.

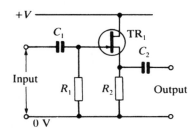

Figure 17.4   *A source-follower JFET amplifier*

This circuit is a common-drain or source-follower amplifier. It does not produce any voltage gain but is invaluable for connecting a high-impedance source with a low-impedance amplifier. High-impedance capacitance microphones of the electret (p. 105) type frequently have an FET amplifier built in to them, powered by a small dry cell. This matches the very high impedance of these microphones to the medium-impedance amplifier or tape recorder to which they may be connected. When such microphones are used at the ends of long screened cables, the high-impedance of the microphone acts in combination with the capacitance of the screened cable to produce, in effect, a low-pass filter. The filter reduces the treble component of the signal, giving a muffled sound after amplification. This type of distortion is reduced by the use of the FET pre-amplifier in the microphone case, for its medium-impedance output is not affected by cable capacitance to the same extent.

MOSFET amplifiers

A typical MOSFET common-source amplifier is shown in Figure 17.5. The transistor is an *n*-channel enhancement type so it operates with its gate at a positive potential. The bias is obtained from a potential divider as in the amplifier of Figure 17.2. The gate requires virtually no current to bias it. R_1 and R_2 may each be several hundred of kilohms and a third resistor R_3 may be placed between the potential divider R_1/R_2 and the gate. R_3 may have a

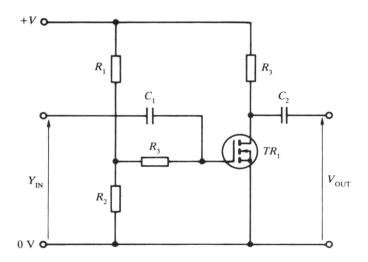

Figure 17.5   *A common-source MOSFET amplifier*

resistance as high as 10 MΩ, so the input impedance of the amplifier is in excess of 10 MΩ, which is very high indeed.

Low-power MOSFETs have a rapid response time, making MOSFET amplifiers useful in radio-frequency circuits. Amplifiers based on power MOSFETs (such as VMOS and HEXFETs, p. 79) are recommended for power-control circuits. They have greater thermal stability than bipolar transistors and are less subject to thermal runaway (p. 65). They may be wired in parallel to enable very large currents to be controlled, in fact a HEXFET really consists of numerous paralleled MOSFETs.

18
Operational amplifiers

The original operational amplifiers were built from discrete components but, since the development of integration, all operational amplifiers (or op amps, as they are more often called) are in the form of integrated circuits. There are dozens of different types, some for general use, others intended for specific purposes, but all operate according to the same principles.

Op amps are not the only type of integrated amplifier obtainable. There is a whole range of IC amplifiers manufactured for special purposes, including audio amplifiers (mono and stereo, pre-amplifiers and power amplifiers), VHF amplifiers, UHF amplifiers, video amplifiers, and pre-amplifiers for infra-red sensors. Each of these has its own special features of concern to the person who is using them, but not of general interest. Op amps and their features are of general interest and such wide application, that they deserve a whole chapter to themselves.

The features of an op amp are as follows. They all have two inputs, known as the **inverting input** (marked '−' in Figure 18.1) and the **non-inverting input** (marked '+'). They operate relative to the 0 V line but they usually take their power supply from a positive line (say, +9 V) and a negative line of equal but opposite potential (−9 V). The connections to the power lines are omitted in Figures 18.1 to 18.3. Also omitted are connections to certain terminals such as the offset null pins. These are provided on the IC to allow the output of the amplifier to be set to zero when both inputs are at the same potential. In other words, to allow for tolerances in manufacture. With many present day amplifiers offset errors are extremely small and the terminals for adjustments are not provided. One important point about op amps is that the inputs have very high input impedance (p. 169). Those with bipolar transistors at the inputs have an input impedance of about 2 MΩ. Many types of op amp have FET or MOSFET input stages. For the latter the input impedance is 10^{12} Ω, or a million megohms. This is virtually infinite input impedance. The output impedance of op amps is usually very low, of the order of 75 Ω.

The gain of an op amp is very high, typically 100 000. This is known as the **open loop gain**. Although we sometimes make use of this high gain, the op amp is more often connected so that the gain of the circuit (as opposed to the gain of the op amp itself), the **closed loop gain**, is less than this. We return to this point later.

175

An op amp is a differential amplifier. Its output depends on the difference between the potentials at the two inputs. Three rules apply:

1   If the inputs are at equal potential, the output is stable. Note that all potentials are measured with respect to the 0 V (ground) line.
2   If the non-inverting input is at a higher potential than the inverting input, the output swings in the positive direction.
3   If the inverting input is at a higher potential than the non-inverting input, the output swings in the negative direction.

Remember that all potentials applied at the inputs and all output potentials can never be greater than the positive supply voltage or less than the negative supply voltage. In some types of op amp, output voltages have a more restricted range and can not lie within a volt or two of the supply voltages.

Using op amps

There are various ways in which op amps are used. Figure 18.1 shows one being used as an **inverting amplifier**. It has two resistors, R_1, the input resistor, and R_2 the feedback resistor. The non-inverting input is connected to the 0 V line. Rule 2 above tells us that, if a positive voltage V_{IN} is applied to the inverting input through terminal A and R_1, the output V_{OUT} swings negative. Point A is positive and point C is negative so somewhere in between there must be a point which is at 0 V. The output will stabilize if the 0 V point is at B, for then the inputs of the op amp are at equal potential (0 V). Under these conditions a current I flows from A, through R_1 to B, the inverting input. We have already said that this has extremely high input impedance so virtually no current flows into B. So I continues unchanged through R_2 to the output C. Along R_1 the potential drop is V_{IN} and:

$$I = \frac{V_{IN}}{R_1}$$

$$V_{OUT} = -V_{IN}\frac{R_2}{R_1}$$

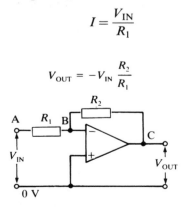

Figure 18.1   *An inverting amplifier circuit, using an operational amplifier*

Along R_2 the potential drop is V_{OUT} and:

$$I = \frac{-V_{OUT}}{R_2}$$

The negative sign is needed because V_{OUT} is negative. Since the current I is the same in both equations we can put their right-hand sides equal to each other:

$$\frac{V_{IN}}{R_1} = \frac{-V_{OUT}}{R_2}$$

$$V_{OUT} = \frac{-V_{IN} \times R_2}{R_1}$$

The final equation shows that the gain of the circuit is $-R_2/R_1$. Note that it depends only on the values of the two resistors, and not on the gain of the actual op amp used. For example, if we make $R_2 = 2.2\,M\Omega$ and $R_1 = 10\,k\Omega$, the gain is $2\,200\,000/10\,000 = 220$. By choosing suitable resistor values we can obtain any required gain that it not as great as the open loop gain of the amplifier.

In Figure 18.1 the current flowing along R_1 is entirely diverted along R_2. Although the inverting input has high impedance it is always automatically brought to $0\,V$ by appropriate voltage changes at the output. The input behaves as if it were an 'earth', that is to say, any current arriving at the input 'disappears' (along R_2) just as though point B was connected directly to earth. We call the input a **virtual earth** in this circuit. Since current entering terminal A is effectively directed to earth at the virtual earth, the input impedance of this circuit equals R_1.

We make more use of the virtual earth in the circuit of Figure 18.2, which is an **adder**. The four (in this case) quantities to be added are represented by different potentials, V_1, V_2, V_3, and V_4. If these are applied to the four inputs of the adder circuit, currents proportional to these potentials flow along the equal input resistors. No current can enter or leave the inverting input so the combined currents all flow along the feedback resistor (also equal to R). To

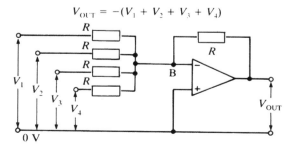

Figure 18.2 *An adder circuit, using an operational amplifier*

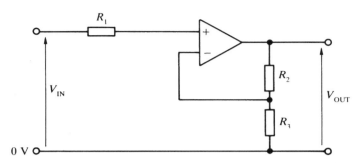

Figure 18.3   *A non-inverting amplifier circuit, based on an operational amplifier IC*

bring the potential at B to zero the output must fall by an amount equal to the sum of the four input potentials. This principle may be used in a circuit in which the output is to be obtained by summing two or more values measured by sensors. It is also used in audio equipment as a mixer, summing signals from two or more sources, such as two tape players. By using variable resistors instead of fixed resistors it is possible to weight the sum, in effect multiplying the signals by different factors to mix in more or less of each signal.

Figure 18.3 shows a **non-inverting amplifier**, with the input connected through R_1 to the non-inverting input of the op amp. R_1 is needed in high-precision circuits but generally it makes little difference if R_1 is omitted. Only a minute current flows into the high-impedance input and therefore there is a negligible p.d. across it. With a given potential V_{IN} at the non-inverting input, the amplifier output will stabilize when this same potential is present at the inverting input. This is the potential at the junction of R_2 and R_3. These two resistors may be considered as a potential divider with V_{OUT} applied across and producing a potential equal to V_{IN}. From the equation on p. 58:

$$V_{IN} = V_{OUT} \times \frac{R_3}{(R_2 + R_3)}$$

Rearranging this equation gives:

$$V_{OUT} = V_{IN} \times \frac{(R_2 + R_3)}{R_3}$$

For example, if $R_2 = 220\,k\Omega$ and $R_3 = 10\,k\Omega$, the gain is $(220\,000 + 10\,000)/10\,000 = 23$. Note that the gain is positive as this is a non-inverting amplifier. This configuration of the op amp is used when inversion of the signal is undesirable. One particular advantage of the circuit is that it has a very high input impedance, the impedance of the op amp input.

In Figure 18.4 the input to the circuit is connected directly to the non-inverting input. We can consider this circuit to be a non-inverting amplifier

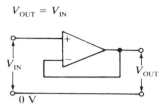

Figure 18.4 *A unity-gain voltage follower circuit, using an operational amplifier*

(see Figure 18.3) in which R_2 is zero and R_3 is infinite. As R_3 approaches infinity, the expression $(R_2 + R_3)/R_3$ approaches 1, or unity. Thus the equation for this circuit is:

$$V_{OUT} = V_{IN}$$

Whatever potential is applied as V_{IN} (within limits) the same potential appears as V_{OUT}. We call this circuit a **unity gain voltage follower**. This circuit is excellent for impedance matching (p. 170). Its input impedance is that of the op amp input, which is at least $2\,M\Omega$, and more if the op amp has FET inputs (p. 76). On the other hand, the output impedance of the op amp is very low.

Instrumentation amplifiers

An instrumentation amplifier (or 'in-amp') can be built from three op amps connected as in Figure 18.5, but it is more convenient and reliable to use an in-amp consisting of three op amps ready connected on a single chip. The in-amp is a differential amplifier which amplifies the potential difference between its inputs. The inputs go directly to the inputs of two of the constituent op-amps, so both inputs are of high impedance and are equal in value, typically 10^9 to 10^{12} ohms. This is not the case in a differential amplifier built from a single op amp, where the impedances are unequal and relatively low. Another advantage of the in-amp is that all or nearly all of the resistors are fabricated on the same chip, so that they are more likely to be accurately balanced.

In Figure 18.5, R_G may be an external resistor connected across two pins of the IC. If the pins are unconnected, the gain of the amplifier is 1 but different gains may be obtained by connecting a resistor of suitable value at this point. Alternatively, a number of resistors may be provided on the chip to give gains of 1, 10, 100 and 1000; one of these can be selected by connecting appropriate pins together.

In-amps are used where long-term stability and high sensitivity are essential. Applications include medical instrumentation such as ECG and EEG circuits.

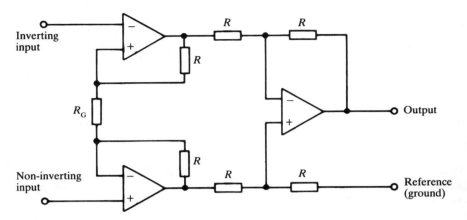

Figure 18.5   *An instrumentation amplifier, consisting of three operational amplifiers; all resistances except R_G have equal values*

There is not enough space here to describe any more of the applications of op amps. As well as their obvious uses as amplifiers, buffers and level detectors, these versatile devices have uses as subtractors, integrators, differentiators, current-to-voltage converters, oscillators and precision rectifiers. At the end of this chapter we look at their uses in active filters. Op amps illustrate so well the way in which integrated circuits have become the building-blocks of electronics.

Transconductance amplifiers

A transconductance amplifier is similar to an op amp, having the same pair of inverting and non-inverting inputs, and the same positive and negative power supplies. It is in fact a type of op amp but with one important distinction, it is the output *current* that is controlled. The output current is proportional to the potential difference between the input terminals. In addition there is a bias input which may be used to switch the output current on or off, or to vary the transconductance of the amplifier.

Transconductance

The voltage gain of an amplifier is the ratio V_{OUT}/V_{IN}. Similarly the current gain of an amplifier is the ratio I_{OUT}/I_{IN}. Both of these are ratios, so they are pure numbers with no units. For a transconductance amplifier the equivalent to gain is I_{OUT}/V_{IN}. This is obtained by dividing a current by a voltage, so it is not a ratio. When we calculate resistance we divide volts by amps (p. 20). In calculating the 'gain' of a transconductance amplifier we divide amps by volts. This is the inverse operation and has inverse units. Instead of *resistance* with *ohms* as the unit, we have *conductance* with *siemens* as the unit. Because the volts are at the input side of the amplifier and the current is at the output side, this is known as *transconductance*.

The transconductance amplifier is often used as an attenuator to control the amplitude of a signal. The signal is fed into the non-inverting input and, according to the potential supplied to the amplifier bias input, the amount of current flowing from the output can be adjusted. This may be fed directly to the base of a transistor for current amplification. A similar application is as an envelope shaper in electronic musical instruments. A piano, for example has a characteristic percussive sound. The note begins with high volume then dies away slowly. Its amplitude envelope shows a sharp rise (attack phase) and a slow fall (decay phase). A note of constant amplitude may be passed from a signal generator, through a transconductance amplifier, to have its envelope shaped to imitate that of a piano. By contrast, the note from a violin usually has a slow attack phase and often an equally slow decay phase. Envelope shaping adds greatly to the realism of electronically generated instrument sounds.

Amplifier gain

When engineers refer to the gain of an amplifier they generally express it in decibels (symbol dB). As explained in the box, gain is a pure ratio with no units. The decibel is not a unit of gain but a scale on which the gain ratio may be expressed. Given two quantities such as V_{OUT} and V_{IN}, the gain is the ratio between them, V_{OUT}/V_{IN}. In decibels, this ratio is expressed by taking its logarithm (to base 10) and multiplying by 10. For example if $V_{OUT} = 4\,V$ and $V_{IN} = 0.5\,V$, the gain is $4/0.5 = 8$. In decibels this is:

$$10 \times \log_{10} 8 = 9 \text{ dB}$$

The decibel scale is a logarithmic one, doubling the gain does not double the decibel value. A gain of 16, for example is equivalent to 12 dB, not 18 dB. Multiplying gain by 100, to make it 800, increases the decibel rating only to 29 dB. Although this scale is difficult to comprehend at first, it is a useful one for the audio designer. It is used not only for expressing gain but the ratio between any two quantities that are measured in the same unit. For example, we may express the ratio between mean signal level and mean noise level in terms of the ratio of their amplitudes and then convert this to decibels. It has the advantage that in a multi-stage circuit the decibel gains of each stage may be added together instead of being multiplied. For example, if an amplifier with a gain of 50 dB is cascaded with an amplifier with a gain of 100 dB, the total gain is 150 dB. The box on p. 184 illustrates the usefulness of the decibel scale.

Power ratios

Ohm's Law tells us that $R = V/I$. Rearranging this equation gives $I = V/R$. The power being dissipated in a resistance is defined as $P = IV$. Combining the equations gives $P = V/R \times V = V^2/R$. Power is proportional to voltage squared. When we are more concerned with the power of the input and output signals of an amplifier rather than the voltage, we take the logarithm of the ratio of the mean input and output voltages and multiply by 20, instead of by 10. Doubling a logarithm has the effect of squaring, so this is equivalent to squaring the voltages *before* finding their ratio. It gives the ratio between output and input *power*, in decibels.

Active filters

The filters described on p. 61 are known as passive filters because they are built from passive components such as resistors, capacitors and inductors. An op amp is an active device, so filters based on op amps are known as active filters. A simple example is shown in Figure 18.6. Its active part is an op amp connected as a non-inverting amplifier (p. 178). By comparison with Figure 6.5a, it can be seen that this filter has two low-pass passive filters in cascade. The two

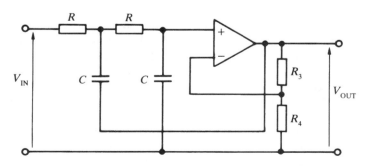

Figure 18.6 *An active second-order low-pass filter, based on an operational amplifier IC*

resistors marked R are equal in value and so are the two capacitors. The signal thus passes through two low-pass filters on its way to the op amp, which is why this is called a second-order filter. The op amp amplifies the signal that reaches it; this is an improvement since, in a series of cascaded passive filters, the output always has a much lower amplitude than the input. But this is not the only benefit. Examination of Figure 18.6 shows that while the second filter has its capacitor grounded on the 0 V line, the first stage has its filter connected to the output of the op amp. The output signal is being fed back to an earlier stage of the circuit. It is a property of a single-stage passive filter (as in Figure 6.5a) that at the cut-off frequency the output signal lags an eighth of a cycle behind the input signal. Two stages of filtering give two eighths of a cycle of delay, or a quarter of a cycle. This delayed signal is fed back to the input stage and acts to reduce the amplitude. At increasing frequencies the delay becomes greater and greater and the resulting output is more and more reduced. At very high frequencies the delay reaches a half-cycle. The input and output are out of phase. This means that when the input signal is going positive the output signal is going negative. When the input signal is going negative, the output is going positive. As a result of this half-cycle delay, the fed-back signal tends to damp out the incoming signal, thus reducing its amplitude to a very low level. High frequency signals are almost completely eliminated.

The action of the filter is complicated but it can be illustrated by a graph (Figure 18.7) on which we plot amplitude against frequency. The frequency is

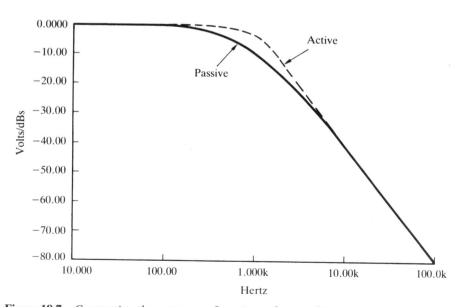

Figure 18.7 *Contrasting the responses of passive and active filters*

plotted on a logarithmic scale, on which equally spaced graduations correspond to a ten-fold increase in frequency. This allows a wide frequency range to be shown on a single graph. The amplitude in volts is plotted in decibels, with respect to the input signal. To start with, study the continuous line, which shows the response of a second-order passive filter. The values of the capacitors and resistors have been selected to give a cut-off frequency of 1 kHz. The output at low frequencies is 0 dB, that is to say it has the same amplitude as the input signal. Above 100 Hz output begins to fall. It falls to about −10 dB at the cut-off frequency, then continues falling until it eventually falls steeply at a steady rate. Measurements on the graph show that the rate of fall is −12 dB per octave. This means a fall of −12 dB for every doubling of frequency.

Using decibels

Figure 18.7 is a good example of the way in which decibels are often a more convenient way of expressing ratios. The input signal to both filters has an amplitude of 1 V. The output of the passive filter has an amplitude of 1 V for low frequencies so, for the passive filter, 0 dB corresponds to 1 V. But the active filter has slight gain. At low frequencies its output is 1.59 V. For the active filter, 0 dB corresponds to 1.59 V. It is difficult to compare the two output curves if we plot them on an absolute voltage scale. Plotting them on a decibel scale makes it easy to see precisely how the output responses differ.

The output of the active filter, with the same resistors and capacitors but with the addition of an op amp, plus R_3 and R_4, is shown by the dashed line. This curve has a much sharper bend in it than the passive curve. The filter passes frequencies just below the cut-off frequency much more readily than does the passive filter. The curve reaches its maximum rate of fall (−12 dB per octave) much sooner. In other words the graph shows that the passive filter has a much narrower zone between those low frequencies that it passes readily and the higher frequencies that it attenuates. It is a much more effective filter. At the cut-off frequency, the signal is reduced to −3 dB, which corresponds to a reduction in signal power of exactly 50%.

The op amp has sharpened the response of the filter. Another way to do this is to include an inductor in the filter circuit, but the disadvantage of this is that inductors of suitable value may be both large and weighty. Also, unless carefully shielded, inductors tend to disseminate and pick up electromagnetic interference. This makes them unsuitable for use in modern compact and portable equipment.

19

Logic circuits

Whether we are reading the time on a digital watch, dialling a call on a cell-phone, listening to a compact disc, using a Smartcard, or sending a fax, we are using logic circuits. More and more of the equipment we use at home and at work depend on electronic logic. Logic is the science of reasoning, and we use its electronic equivalent as the basis for a whole host of so-called 'intelligent' devices. The form of logic most suitable for implementing electronically is **binary logic**. In binary logic, we constantly deal with two states. A statement is true or it is untrue: there are no half-truths. A digit is 1 or 0: there are no fractions or other values. A transistor is either fully on (saturated) or fully off: it switches rapidly from one state to the other, spending a negligible time in the intermediate states. A voltage is either high or low: intermediate values have no meaning.

The on-off, high-low characteristics of binary logic mean that we can design electronic circuits that can model logical statements with absolute accuracy. We need to be concerned with only two voltage levels, not with all the possible levels that exist in (for example) audio circuits. Let us see how a logical situation may be modelled electronically. A garage floodlamp is to be switched on whenever a car approaches the garage, but this is to happen only at night. There is no point in turning on the floodlight during the day. There are two sensors. One of these, perhaps based on a light-dependent resistor (p. 120) responds to light level. It can be arranged, possibly by using a Schmitt trigger circuit (p. 96), that the output voltage of the sensor circuit is low (close to 0 V) when it is daylight and high (close to the supply voltage) at night. The other sensor is a pyroelectric sensor; it responds to the heat radiating from the car. Its circuit is arranged to give a high output when it is triggered.

The logic required is simple. If the LDR detects night AND the pyroelectric device detects a car, the floodlight comes on. The AND in this sentence is an example of a logical operator. It links the two conditions under which the floodlamp is to be turned on. It tells us that we must have one condition (night) AND the other condition (approaching car) in order to turn on the floodlamp. Putting it another way, we have three statements:

A = it is night
B = a car is approaching
Z = the floodlamp is on

Any of these statements may be true or not true, but the truth or otherwise of Z depends on the truth of A and B. The logic of the lighting control, expressed in short form is:

IF A is true AND B is true THEN Z is true

or even shorter:

IF A AND B THEN Z

The 'IF . . . THEN . . . ' format appears often in binary logic.

Inverse statements

Each of the statements above can be paired with an 'opposite' statement so that, if the statement is not true, its 'opposite' or inverse is true. The inverse is represented by the same letter as the original statement, but with a bar over it. For example, if A = 'it is night', then $\bar{A}$ = 'it is not night'. In most contexts, including the example above, we accept that 'it is not night' means the same as 'it is day'. But, at dawn and dusk, people could be justified in stating that it is not night, yet it is not day either. So, 'it is day' is not a strict inverse of 'it is night' The only exact inverse is 'it is not night'. This difficulty does not arise with Z = 'the floodlamp is on' ,which has the inverses $\bar{Z}$ = 'the floodlamp is not on', and $\bar{Z}$ = 'the floodlamp is off' because 'off' and 'on' are exclusive opposites.

Logical operators

There are three basic logical operators that are used in 'IF . . . THEN . . . ' statements:

NOT: IF A is true, THEN $\bar{A}$ is NOT true. Since all statements must be true or not true, this implies that if $\bar{A}$ is NOT true THEN A is true.
Example IF it is night THEN it is NOT daytime (assuming we ignore dawn and dusk).

AND: IF A is true AND B is true, THEN Z is true. Otherwise Z is not true (false).
Example IF it is night AND a car approaches, THEN the floodlamp comes on.

OR: IF A is true OR B is true, THEN Z is true. Otherwise Z is not true (false).
Example IF it is night, OR a car approaches, THEN the floodlamp comes on.

The full operation of the floodlamp controller depends on the AND operator. We shall shortly look at ways in which this operation can be performed electronically.

Truth tables

The action of logical operators can be represented by truth tables in which '1' represents 'true' and '0' represents 'not true', or 'false'. We list the conditions A, B under 'Input' and the consequence Z under 'Output':

NOT			AND			OR		
Input A	Output Z		Input A	B	Output Z		Input A	B	Output Z
0	1		0	0	0		0	0	0
1	0		0	1	0		0	1	1
			1	0	0		1	0	1
			1	1	1		1	1	1

Logical statements can be extended to include more than two conditions. For example, IF A is true AND B is true AND C is true AND D is true, THEN Z is true. Otherwise, Z is false.

It is a simple matter to perform the basic operations using ordinary manually operated switches, with a lamp to indicate the logical outcome. Figure 19.1a shows how we perform the AND operation on A, B, C and D. Figure 19.1b shows the OR operation. For calculators, clocks and computers we need full electronic switching at high speed, combined with reliable action, freedom from interference from occasional voltage fluctuations and 'spikes', and outputs that are firmly high or low. Logic circuits that meet these specifications are made in integrated form and we consider some of these in the next section.

All logical situations can be expressed in terms of NOT, AND and OR, but there are three other operators that are often used to simplify the statements.

EXCLUSIVE-OR, sometimes known as EX-OR is a variation on OR. In the OR truth table we can see that Z is true when A OR B OR BOTH are true. In EXCLUSIVE-OR Z is true when A OR B but NOT BOTH are true. Another useful operator is **NOT-AND**, which is the AND operation followed by NOT. This is usually shortened to **NAND**. Similarly, there is **NOR**, which is OR followed by NOT. The truth tables of these operators are:

EXCLUSIVE-OR			NAND			NOR		
Input		Output	Input		Output	Input		Output
A	B	Z	A	B	Z	A	B	Z
0	0	0	0	0	1	0	0	1
0	1	1	0	1	1	0	1	0
1	0	1	1	0	1	1	0	0
1	1	0	1	1	0	1	1	0

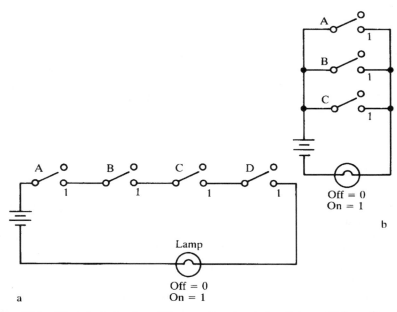

Figure 19.1   *Electrical circuits with manually operated switches, which perform logical operations (a) AND (b) OR*

All logical operations can be performed using just the NAND and NOR operators, though the statements using only NAND and NOR may be more difficult to understand that those which use the simpler AND, OR, and NOT operators.

Transistor-transistor logic

Although transistor-transistor logic (or **TTL**, as it is more usually called) has been largely superseded by other forms with superior performance, it is TTL which was the dominant type in the expanding days of the seventies and early eighties. Many of the logic ICs today are simply upgraded versions of the original TTL, performing the same functions and having the same pin connections.

The circuit element that performs a single logical operation is known as a **gate**. Figure 19.2 shows the circuit of the TTL gate that performs the NAND operation. The circuit has an unfamiliar component which is a transistor with two emitters. If both emitters are connected to the positive supply voltage (logical high, equivalent to A AND B) the transistor does not conduct. The effect of this on the remainder of the circuit is to make the output Z go low. The transistor is turned on when either one or both emitters are connected to 0 V (logical low), any combination of inputs making the output go high. Thus the behaviour of the circuit mirrors the action of the NAND operator. One might wonder why it requires such a complicated circuit to perform this operation. One function of the other components is to ensure fast action. The other function is to ensure that the voltages at high and low levels are clearly distinguished, with no half-way states possible.

Even though the circuits of the descendants of TTL are appreciably more complicated than Figure 19.2, they are still simple enough from the point of

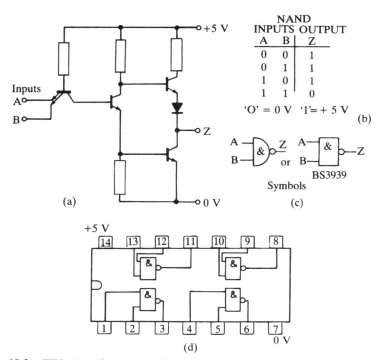

Figure 19.2 *TTL (a) The circuit of a two-input NAND gate (b) Its truth table (c) Symbols for NAND gates (d) The 7400 IC which contains four two-input NAND gates, with shared power-supply pins*

view of integration, and they occupy little space on the silicon chip. Four independent gates are fabricated on one chip in a 14-pin package as shown in Figure 19.2d.

TTL and its derivatives comprises a large family of several hundred different ICs. Some of these consist of relatively simple gates but others employ MSI and LSI (p. 142) to build complex counters, registers and logical arrays on a single chip. The chief advantage of TTL is its high speed, making it suitable for use in computers. It requires a supply voltage between 4.75 V and 5.25 V so a regulated supply is virtually essential. Electrical noise such as 'spikes' on the supply lines are liable to upset its operation. Its main disadvantage is that it requires fairly large currents to operate it. Even a relatively small system consisting of half-a-dozen ICs may require a supply of 1 A, which means that the use of TTL in battery-powered equipment is severely limited. This has prompted the development of versions of TTL that are less power-hungry. One of the most popular of these is low-power Schottky TTL, or LSTTL. Power requirements are reduced to about a fifth of those of standard TTL by increasing the values of the resistors in the circuits. The name 'Schottky' refers to an element in the circuit known as a Schottky clamp. The function of this is to prevent transistors that are turned on from being fully saturated. Saturated transistors take longer to switch off than transistors that are not saturated because it is first necessary to remove excess charge carriers from the base region of the transistor. The Schottky clamp forward-biases the transistor by only a few tenths of a volt so that it turns off quickly. The LS series is not much faster than standard TTL, but the newer Advanced low-power Schottky version is twice as fast and uses about one-twentieth of the power of standard TTL.

CMOS logic

Although standard CMOS logic (known as the 4000 series) operates about 10 times more slowly than TTL and its variants, this is no disadvantage in many applications. Its chief advantage is that it has very low power consumption, making it ideal for battery-powered portable equipment. When gates are not actually changing state they require practically no current at all, so quiescent power requirements are very low indeed. CMOS operates on any supply voltage between 3 V and 15 V. A regulated power supply is not required. An LSTTL output can drive only about five LSTTL inputs and similar limitations apply to the other TTL versions. By contrast a CMOS output can drive a practically unlimited number of CMOS inputs. This makes circuit design much simpler.

The 'C' in the acronym CMOS stands for 'complementary' and Figure 19.3 explains what this means. The gate consists of two enhancement-mode transistors of complementary type, n-channel and p-channel. Given a high input

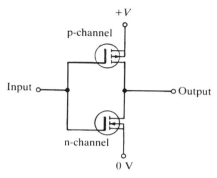

Figure 19.3 *A complementary MOS gate that performs the logical NOT operation*

the p-channel transistor is turned off and the n-channel input is turned on. This more or less short-circuits the output to the 0 V rail, giving a low output. Conversely, a low input turns on the p-channel transistor and turns off the n-channel transistor, short-circuiting the output to the positive supply rail. The action corresponds to the logical NOT operator. The NAND gate illustrated in Figure 19.4 operates in a similar fashion.

Because CMOS works by short-circuiting the output terminal to one or other of the power rails, the output voltage is very close to one or other of the supply rails. There is a clear distinction between the high output level and the low level. This is why CMOS is relatively unaffected by noise and 'spikes' on its supply lines. Figure 19.4 also illustrates the fact that the CMOS gate requires few components. Thousands of gates can be accommodated on a

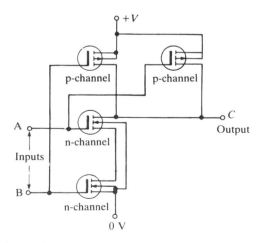

Figure 19.4 *The circuit of a CMOS two-input NAND gate*

single chip so that CMOS is ideal for LSI and VLSI. Consequently the CMOS range of logic circuits includes devices such as 16-stage counters and 64-stage shift registers that are not available in TTL.

CMOS transistors, like all MOS transistors (p. 78) have very high input impedance, so they require virtually no current to alter the potential of the gate. We can understand why the current requirements of CMOS are exceedingly small, and why current is required only when the input to a gate is being changed.

Transmission gates

Figure 19.5 shows a CMOS gate in which the n-channel and p-channel transistors are both switched on by a high control input and are both switched off by a low control input. The circuit acts as an electronically-controlled switch and can be controlled by the output from another CMOS gate. When the transistors are off, only a minute current of a few nanoamps can flow between the terminal on the left to the terminal on the right. When the transistors are on, the resistance between input and output terminals is only about $80\,\Omega$, the gate acting as a low-value resistor. Current can flow freely in either direction, which is why we refer to these as input/output terminals. Analogue signals (for example, audio signals) can be transmitted in either direction.

The gate in Figure 19.5 acts as a single-pole on-off switch. Other CMOS transmission gates are made which consist of two or more transmission gates with common control lines. These make it possible to implement other switching actions, such as change-over switches, double-pole double-throw switches, and 1-of-4 (or more) selector switches, all under logical control.

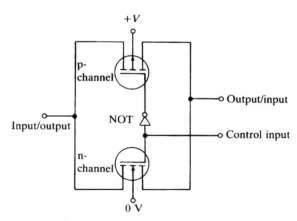

Figure 19.5   *The circuit of a CMOS transmission gate*

Other logic

There are several other logic families each with its own characteristics and special applications. One such family is high-threshold logic (HTL) or high noise immunity logic (HNIL) in which the gap between the low and high logic levels is particularly large. This makes the series suited to industrial environments where electrical noise would make it impractical to use TTL or CMOS. Emitter-coupled logic (ECL) is designed for very high-speed operation; as with Schottky devices the transistors are prevented from becoming completely saturated, but in this family this is achieved by have the logical low and high levels close together. This makes ECL liable to noise interference and also requires more power as the transistors are bipolar types and are always switched on. One very popular family is the HC family which has CMOS-based gates but runs at high speed. It combines the best features of TTL and CMOS. Most of the standard TTL devices are available as HC types, and there are HCT devices intended for interfacing directly with TTL.

Logic and numbers

Electronic logic involves only two states; on and off, high and low, 1 and 0. But the numbers we use in everyday life rely on having 10 different symbols (0 to 9) to express them. To convert these directly to electronic form would mean having 10 input and output levels. Although it is possible to do this, the circuitry required is much more complicated than that used for logic gates. We throw away the advantages of high-speed operation, reliability, noise-immunity, and the ease of implementing the circuits in MSI and LSI. It is much more sense to abandon the decimal system and replace it with a number system that has only two values, 0 and 1. This is known as the **binary system**.
    Counting in the binary system runs like this:

Decimal number	Binary equivalent
0	0
1	1
2	10
3	11
4	100
5	101
6	110
7	111
8	1000
9	1001
10	1010

In the decimal system, the value of a digit is determined by its place in the number. For example, the number '374' means 3 hundreds plus 7 tens plus 4 units. In the number 6267, the first 6 means 6 thousands, but the second 6 means 6 tens. Value depends on position. As we go from right to left, the position value increases by 10 times.

The binary system works on the same principle but in twos instead of tens. The position values from right to left are units, twos, fours, eights, sixteens and so on, doubling up each time we shift one place to the left. So the binary number 110 is the equivalent in decimal of 1 four plus 1 two plus no units, totalling $4 + 2 + 0 = 6$. Similarly, the binary number 1010 is the equivalent of 1 eight plus no fours plus 1 two plus no units $= 8 + 0 + 2 + 0 = 10$.

Bits and bytes

The short word for a binary digit is **bit**. Thus the number 101 has three bits. The bit on the right, the unit bit, is known as the **least significant bit**. The bit on the left, the 'fours' bit is known as the **most significant bit**.

A string of eight bits is known as a **byte**.

Mathematics is a logical process so it is not surprising that we are able to use logic circuits to perform mathematical operations. As a simple illustration, consider the addition of two single-digit binary numbers, either of which may be 0 or 1. The results of all possible additions with these two numbers are:

INPUT		OUTPUT	
A	B	Sum, S	Carry, C
0	0	0	0
0	1	1	0
1	0	1	0
1	1	0	1

The addition in the last line would normally be written with the carry digit on the left: $1 + 1 = 10$. In decimal this is the equivalent of $2 + 2 = 4$ as can be seen from the table above. If we compare the value S in the sum column with the output values of the truth tables on p. 188 we find that S is the equivalent of Z for an EXCLUSIVE-OR gate. Similarly, comparing the value C in the carry column we find that this is the same as Z for an AND gate. Given an EXCLUSIVE-OR gate and an AND gate, we can use them to add two single-digit binary numbers (Figure 19.6).

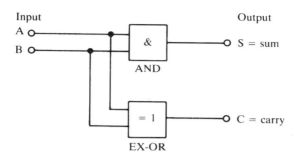

Input   Output

A   &   S = sum

B   AND

            = 1   C = carry

           EX-OR

Figure 19.6   *Using logic gates to add two single-digit binary numbers*

Addition of binary numbers with more than a single digit is naturally more complicated, as we have to allow for adding in the carry digits at each stage, but it is easily done with suitably connected logic gates. The addition circuit is usually presented as an MSI device. Other mathematical operations can also be performed by logic. For example, we can multiply two numbers together by adding the same number to itself repeatedly and counting the number of times we add it.

Programmable logic

Logic circuits can be built from a standard range of logic ICs of the kinds described above. All we need to do is to work out what gates are required and how to connect them together. Anyone who has attempted to design even a fairly simple logic circuit knows that the number of ICs required can easily mount to the twenties, thirties or even forties. These need a large circuit board, consume a large amount of power and require a complicated pattern of inter-connections on the board. For a mass-produced circuit of some complexity, such as a computer, it is possible to have specially designed LSI or VLSI circuits incorporating all the logic on one chip. But setting up production of a special IC is expensive and unlikely to be economic except for very large runs. Also, any change or improvement in the design requires the masks to be scrapped and re-designed. With programmable logic we use ICs that are pro-duced in a few standard types, and so are inexpensive, but can be programmed to perform any of a wide range of logical operations.

One of the simplest forms of programmable logic IC is programmable read-only memory (PROM). These are normally used to store computer programs in a permanent form: for example, the system program that a computer uses when it is first switched on, to get it running. As an example, consider the simple PROM shown in Figure 19.7a, which has fewer inputs and outputs than

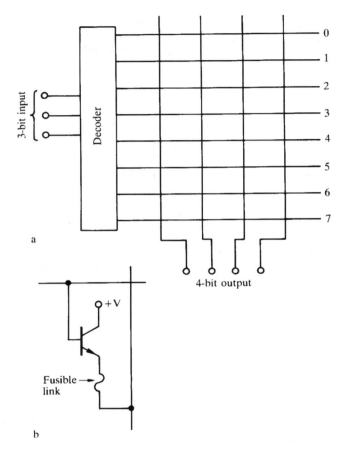

Figure 19.7 *The principle of a programmable read-only memory (a) Simplified schematic diagram (b) The transistor and fusible link that joins lines 0–7 to the output lines at every crossing-point in (a)*

is usual but serves to illustrate the principles. This has three inputs and four outputs. The inputs go to a three-line to eight-line decoder. For each of the eight possible combinations of states of the inputs (000 to 111), one of the eight lines goes high. There is a link (Figure 19.7b) between each of these lines and the four outputs, the link consisting of a transistor and a fuse. When the device is programmed, other circuits not shown in the figure, allow us to decide which fuses are to be intentionally blown and which are left intact. This is what we mean by saying that the device is programmable. Suppose the input is 110, so that line 6 is made high. If all the fuses are present, base current is supplied to all transistors on that line and all four output lines go high; the output is 1111.

But some of the fuses may have been blown when the device was programmed. If any fuse is blown the corresponding output stays low.

Depending on which fuses are blown, the output corresponding to an input 110 can be any one of the sixteen 4-bit binary numbers from 0000 (all blown) to 1111 (none blown). Once a fuse has been blown it can not be repaired, so the device is programmed once and for all. After that it is possible only to read the data stored in it, but not to change it, which is why it is called a **read-only memory**.

Used as a memory, the device is programmed with data which might be a table of information or might be the steps of a program in code. According to the 3-bit input (or address), we obtain an output corresponding to the stored data. The use of the PROM as a logic array is similar, as shown by this example. An automatic lamp-switching system for a shop has three inputs: a passive infra-red detector (p. 105) in the store-room, a light sensor (p. 120) to detect if it is day or night, and a weekly timer to register if the day is Sunday or not. The output goes to four sets of lamps: the shop, an illuminated display in the shop, the store room, and the car park. The lamps are to be switched on as follows:

In the shop: every night, every day of the week.
The display: during day-time, every day, except Sunday.
In the store room: all the time, except day-time on Sunday.
In the car park: if a car is detected at night, except Sunday night.

The truth table for these requirements is:

INPUT			OUTPUT			
I-R detector	*Day/Night*	*Day of week*		*Lamps (1 = On)*		
(1 = car)	(1 = night)	(1 = Sunday)	Shop	Display	Store	Car park
0	0	0	0	1	1	0
0	0	1	0	0	0	0
0	1	0	1	0	1	0
0	1	1	1	0	1	0
1	0	0	0	1	1	0
1	0	1	0	0	0	0
1	1	0	1	0	1	1
1	1	1	1	0	1	0

The table has eight rows, covering all possible combinations of inputs. The outputs could be obtained by using a number of NAND and NOR gates but blowing the fuses of a PROM to correspond with the 1's in the output columns of the truth table achieves the same effect with just a single IC and there is no need for connections between gates.

Sequential logic

The logic we have considered so far is **static logic**. We set up a number of fixed inputs to the circuit and the output takes a given value, depending upon the truth table or tables of the circuit. Given a set of inputs, the circuit always behaves in the same way. In **sequential logic** the behaviour of the circuit is partly or wholly dependent on what has been happening to it previously. A sequential circuit built with transistors is the flip-flop (or bistable) circuit shown in Figure 15.2. We give it a low input by connecting input A briefly to ground. What happens as a result of this input depends upon what state the circuit is in, which depends on the nature of the previous input. If output A is high, nothing happens. If output A is low, it changes to high. The behaviour of the flip-flop goes through a sequence of events. This is just a simple example of a sequential circuit, but from a simple flip-flop we can build up more complicated circuits such as counters. The four outputs from a counter made from four connected flip-flops step through the sequence 0000, 0001, 0010, 0011, . . . , 1111 each time the input to the first flip-flop is grounded. It counts from 0000 to 1111 in binary, that is from 0 to 15 in decimal. Sequential logic is important for the control of equipment such as an automatic drilling machine or a robot which has to perform a sequence of different tasks.

Logic circuits and the outside world

Logic circuits operate on electronically-represented facts but a logic circuit on its own cannot operate on any facts unless it has been told what they are. At the other end of the process, the circuit needs to be able to communicate the results of its operations to the user.

One of the most obvious and simplest ways of giving facts to a logic circuit is to open or close a switch. It might be a micro-switch attached to a door. When the door is open the switch is open; when the door is shut the switch is closed. The switch tells the logic circuit about the state of the door. But even such a simple arrangement has its problems. One of the major ones is **contact bounce**. It happens when we operate any switch, such as a key on a keyboard. As the key is pressed and the contacts come together, they do not necessarily stay together. They may make and break several times as the switch closes, closing together finally when the switch has reached the end of its travel. Contact bouncing may take as long as 20 ms.

This 'off-on-off-on-off-on-off-on' nature of a switch makes no difference when, for example, we are switching on a room lamp. The response of the lamp is slow, since it takes an appreciable time to heat up the filament to the temperature at which it begins to emit light. By the time it is heated the contact bouncing is over. The lamp does not appear to flash on and off. But logic circuits are fast, operating in microseconds or even nanoseconds. They have no difficulty in responding to every off-on change. If a logic circuit is being used to count how many times we press a key it counts once for every off-on change. We press the key once, but the counter may register 4 or more off-on changes.

The solution to this problem is to de-bounce the switch or key. Figure 19.8 shows a debouncing circuit using a flip-flop. Here the flip-flop is a logic IC, not made from discrete components as in Figure 15.2, but its action is the same. The two inputs, corresponding to inputs A and B in Figure 15.2 are known as the 'set' (S) and 'reset' (R) inputs. When not connected to the 0 V line by the switch they are held high by the resistors. Connecting S to 0 V sets the flip-flop. Its output Q goes high. Connecting S to 0 V again has no effect for the flip-flop is already set. Its output can be changed only by connecting R to 0 V, which resets the flip-flop, making Q go low.

In the figure the switch is connecting S to 0 V so the output is high. As we move the switch to its other position, contact bounce will cause makes and breaks between S and the 0 V line, but this has no effect on the flip-flop. But the first time that the switch connects R to the 0 V line, the flip-flop changes state and its output goes high. Having done so, it is completely unaffected by any subsequent makes and breaks between R and 0 V.

One of the most important differences between a logic circuit and the outside world is the way in which quantities are represented. In the real world, temperature is a quantity which changes smoothly as an object is warmed or cooled. On a mercury thermometer we can see the end of the column moving smoothly up or down the tube. The thermometer scale may be marked in degrees but we can estimate fractions of a degree quite easily. The same applies when we use a thermistor (p. 104). Temperature changes smoothly and so does the resistance of the thermistor. If the thermistor circuit represents changes in resistance as changes in voltage at its output, we find that the voltage changes smoothly too.

The way the voltage changes mirrors the way the temperature is changing. The rate and extent of the change in voltage corresponds to the rate and change of the temperature. We say that the voltage is an **analogue** of the temperature.

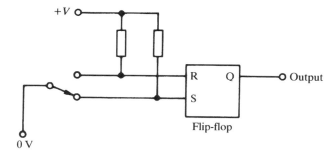

Figure 19.8 *Using a flip-flop to debounce a switch*

In the same way, a microphone is sensitive to pressure changes in the air caused by sound waves. It generates an alternating voltage which we can amplify. The waveform of the amplifier output is an analogue of the waveform of the original sound.

By contrast there are no smoothly varying quantities in a logic circuit. Voltages change abruptly from 0 to 1 and 1 to 0, and in-between states are not recognized. A value is represented as a binary number which exists in the circuit as a series of highs and lows, representing the digits of the binary number. The value is represented in digital form. Informing a logical circuit about quantities in the outside world first involves converting it to an electrical analogue, usually a voltage, and then converting the voltage into digital form. Converting into an analogue is done by using a sensor such as a thermistor or microphone with a suitable circuit such as an amplifier to produce an analogue voltage. Converting this voltage into digital form is done by using an **analogue-to-digital** converter, or ADC.

The details of how an ADC works are too complex to go into here. An ADC is usually an integrated circuit which has a single input to which the analogue voltage is fed. It has a number of logic outputs, often eight or twelve, which may be high or low. The outputs produce an 8-bit or 12-bit binary number, its value depending on the voltage fed to the IC. If there are 8 bits, the output ranges from 0000 0000 to 1111 1111. In decimal this is from 0 to 255. There are 255 steps from the bottom to the top of the range. For convenience (and also to simplify this explanation) we may set the analogue input to range over the values 0 V to 2.55 V. 2.55 V divided into 255 steps gives 0.01 V per step. If the input increases by 0.01 V, the output increases by 1. If the input is 1.47 V, the output is 147 in decimal, represented by 10010011 in binary. Figure 19.9 summarizes the action of an ADC.

With an 8-bit ADC as described above, while the input may vary smoothly, taking any value between 0 and 2.55 V, the output is allowed to take only 256 different values in steps of 0.01 V. The resolution of the ADC is 0.1 V. This may be sufficiently good for many applications. A 12-bit ADC gives much better

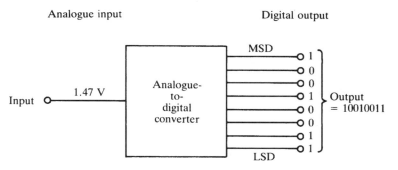

Figure 19.9   *The action of an analogue-to-digital converter*

resolution since there are 4096 steps in increasing from 0000 0000 0000 to 1111 1111 1111.

When a logic circuit has performed its logic and its calculations it needs to make contact with the outside world again. Logic circuits can drive indicators such as LEDs directly, as well as seven-segment LED displays and LCD displays. Through the action of a transistor, a relay or an optocoupler, they can switch lamps, motors, solenoids and various other devices on or off. Sometimes a simple on-off action is not enough. The logic circuit may be required to control the brightness of a lamp or the speed of a motor, for example. In such cases the circuit must be able to produce a variable output voltage. This is usually done by using a **digital to analogue converter**, or DAC. The action of the DAC is the reverse of that of the ADC. It accepts an 8-bit or 12-bit binary input from the logic circuit and produces a voltage output of equivalent value. The output voltage is then used to control the brightness of the lamp, the speed of the motor or any other variable-rate process.

20

Audio electronics

One of the principal aims of audio electronics is the true recording and reproduction of sound. Other aspects of audio electronics include the generation of musical sounds, a topic that is dealt with later in the chapter. In describing recording and reproduction we refer only to musical instruments, but most of what is said applies also to speech and song as well as other sounds and noises. Between the original sound and the reproduced sound there is a chain of processes which is illustrated in Figure 20.1. On the left of the figure is the original sound, a pattern of wave vibrations in the air made by one or more musical instruments or singers. It consists of alternating regions of pressure and rarefaction rippling through the air. The air particles move to and fro, in the same direction as that in which the waves are spreading. We call them **longitudinal waves**. In diagrams, it is easier to draw the waves as if they were transverse waves, like ripples on the surface of a pond.

The first step in the chain is the microphone. Most microphones are transducers (p. 103) converting the energy of motion of the air particles first into the motion of a diaphragm and ultimately into an alternating e.m.f. This e.m.f. is ideally a perfect analogue (p. 199) of the original sound. Its variations should

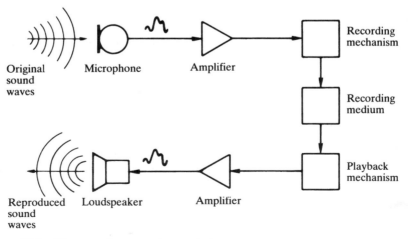

Figure 20.1 *The recording-reproducing chain*

202

exactly characterize the variations in the motion of the air particles. If they do not, we have already lost some of the fidelity that we are striving to attain. This first link in the chain is essentially a mechanical one and its details fall outside the scope of electronics. Some types of microphone have better qualities than others and for most high-fidelity audio work a ribbon microphone is preferred. Other types, such as capacitor and electret microphones, may be preferable in certain circumstances.

Tape recording

We follow the next stages in the audio chain in the context of recording analogue signals on magnetic tape. The magnetic tape recorder became available in the late 1940s and has become very popular. The facility of being able to record at home without requiring additional or expensive equipment has been a big factor in its widespread use. The principle of magnetic tape recording is illustrated in Figure 20.2. The polyester tape is coated with a layer containing a magnetic substance such as ferric oxide, chromium dioxide or pure iron in a finely divided form. The coating can be thought of as partitioned into a large number of minute regions known as domains. In any one domain, all the molecules of the material are arranged in one direction, so the domain is, in effect, a small magnet. In unmagnetized tape the domains are magnetized in randomly different directions (Figure 20.2a) so there is no magnetic field around the tape as a whole. If the tape is fully magnetized (saturated), all domains are magnetized in the same direction (Figure 20.2b). In a tape that carries a recording, an intermediate state exists in which a proportion of domains are aligned but the remainder point in random directions. During recording the tape is moved at a fixed speed past the recording head. This consists of an electromagnet which has a narrow gap between its poles (Figure 20.2d). The signal from the amplifier is fed to the coil, so the magnetic field is an analogue of the original sound. The proportion of domains similarly aligned and the direction in which they are aligned depends on the strength and direction of the magnetic field as the tape passes the recording head (Figure 20.2c).

On playback, the tape passes over a playback head. This is similar to the recording head but the action is the opposite. On all but the most expensive machines the same head is used both for recording and playback. As the tape passes the head the field produced by its magnetization induces a varying current in the coil of the playback head. The varying current is an analogue of the original sound. Recording of stereophonic sound is easy with a tape recorder. It simply needs two heads, one for each channel. The heads are mounted one above the other as a single unit and produce two parallel tracks on the tape.

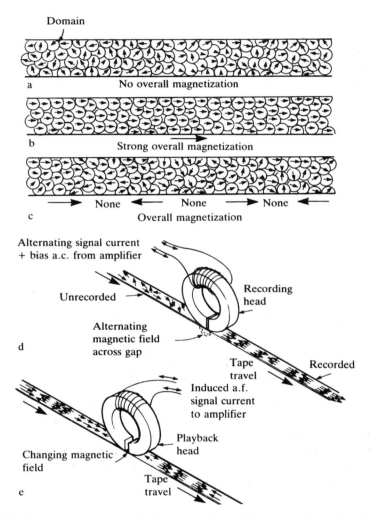

Figure 20.2 *Using magnetic tape (a) Unrecorded tape (b) Saturated tape (c) Recorded tape showing the direction of magnetization of the domains (size greatly exaggerated) (d) Recording (e) Playback*

Induction and motion

Induction occurs only when there is a change of magnetic field. The current induced in the playback head is the result of changes in the magnetic field as the tape passes the gap between the poles. If the tape is stopped, there is no change, no induction and no signal.

When a region of tape is magnetized during recording, the speed of the tape must be such that this region has moved away from the head before any significant reversal of the field occurs. The higher the tape speed, the better the quality of the sound. The standard speed for a cassette tape is 47.625 mm per second. This is half the standard speed previously used on open-reel recorders and is made possible by the improvements in tape and head quality made since the early days of tape recording. For playback it is essential that the head should have a narrow gap (about 1.2 μm) so that it is subjected to only a restricted region of the tape as it passes over the head. If a separate recording head is used, it may have a wider gap, perhaps 10 μm. The fact that the magnetic field may vary in strength and may reverse in direction while a point on the tape crosses the gap does not have any effect on the quality of the recording. This is because the magnetization of the tape depends on the field as the tape leaves the gap. Effective magnetization is confined to a narrow region of tape adjacent to the pole last passed by the tape.

One of the features of magnetic materials is that they do not become magnetized at all unless the magnetic field exceeds a certain minimum strength. This means that signals at low level are not recorded. The audio signal is one that alternates and passes through zero twice during each cycle. Figure 20.3a shows what happens if an unmodified audio signal is applied to the recording head. We are looking at a part of one cycle of the signal as it swings from its negative peak through zero to its positive peak. At both peaks the magnetization is strong (more about that in the box) but when the signal is close to zero the tape is not magnetized. On playback the signal has a step discontinuity as shown in the figure, which is a considerable distortion of its previous smooth form. The problem is overcome by the use of a.c. bias (Figure 20.3b). Before recording it, the audio signal is mixed by a high-frequency a.c. signal. This in effect brings the average strength of the magnetic field up into the range at which it is recordable. Since the a.c. signal is in the ultrasonic range (usually 90 kHz), it is not heard on playback.

Linearity

Magnetization is not proportional to field strength. Their relationship does not give a straight-line graph – it is not linear. At low levels of field strength there is little or no magnetization (which is why bias is needed, Figure 20.3). At high levels the tape approaches saturation. In between low and high there is a region in which magnetization and field strength are very close to being proportional but even in this region there is always a certain amount of distortion of the original waveform.

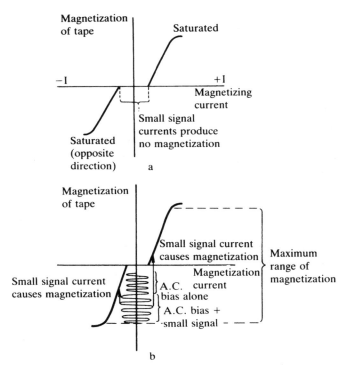

Figure 20.3   *The action of AC bias (a) with no bias; (b) with bias*

Erasing

One of the advantageous features of magnetic tape is that recordings can be erased and the tape used again for other recordings. It also allows corrections to be made to existing recordings. Erasing consists in disarranging the domains, so that they come to lie in random directions, as on an unrecorded tape. The technique is to subject the tape to a strong alternating magnetic field and gradually reduce the strength of the field to zero. Domains vary in the strength of field needed to alter their alignment. When the field is strong, all domains are affected at each change in direction of the field but, as the strength of the field is reduced, more and more domains become unaffected, each remaining in the direction it was in when it first became unaffected. Eventually all are randomly orientated.

Most tape recorders have a special erasing head. This is fed by the same a.c. signal as is used for bias, though at greater strength. The gap of the erasing head is large so that its field extends over a considerable length of tape. As a region of tape passes away from the head it experiences a gradually decreasing alternating field and is demagnetized.

The final stages

Returning to Figure 20.1 we see that the playback of the tape is followed by amplification of the induced signal from the playback head to give it enough power to drive a loudspeaker. The cone of the loudspeaker is made to vibrate and this causes the air around it to vibrate in longitudinal waves, creating sound. The aim has been to make this sound as indistinguishable as possible from the original sound. Just as much depends on the later stages as on the early ones. In particular, the design of the loudspeaker and its enclosure play a major part, but these are mechanical aspects of sound reproduction that we can not go into here.

Similar considerations apply when a recording on one tape is copied on to a new tape. The original recording taken at a live session may be as nearly perfect as is possible. During the production of the final version of the recording there may be several stages in which tracks from different masters are mixed on to one tape. Multiple copies of the final version must be made for selling to the public. There is a loss of quality at every copying stage and the marketed version can never be as good as the original.

Before leaving this topic it should be mentioned that the fidelity of reproduction depends on many factors other than those already mentioned. These include the smoothness of the tape surface, the type of magnetic material used and the construction and materials of the tape head. There are also the consequences of age and wear. Tape stretches with age and its surface may become abraded. The magnetic heads eventually become worn and the gap width increases. These factors lead to progressive deterioration in sound quality.

Digital Audio Tape

Digital audio tape, or DAT, is a relatively new recording technique that overcomes many of the defects of the analogue recording technique described above. The stages in the chain are the same as those shown in Figure 20.1 with the addition of an analogue-to-digital converter (ADC) before the recording stage and a digital-to-analogue converter (DAC) after the playback stage. From recording to playback we operate with a digital signal instead of an analogue signal and this makes a lot of difference to the quality of reproduction.

Analogue-to-digital conversion is described on p. 200. In Figure 20.4 we see this applied to an audio signal. The graph shows the waveform of one cycle of an audio signal. A sample-and-hold circuit registers the signal voltage ten times

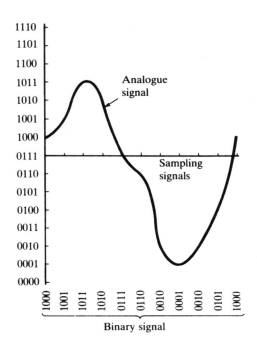

Figure 20.4   *Digital representation of an analogue audio signal*

during each cycle. The sampled analogue value is held long enough for it to be converted into a four-bit binary number by an ADC. On the left of the figure the voltage scale is marked using binary numbers. At the bottom of the figure we see the successive readings, also expressed in binary. This is the digital equivalent of the analogue signal.

If we were to take the eleven values given along the bottom of Figure 20.4 and plot a graph of them, the graph would look like Figure 20.5. This is what the reproduced signal would be like if we fed the binary signal to a digital-to-analogue converter. It is but a crude representation of the original, but there are two ways of improving it. One way is to take samples more often. Instead of sampling only ten times, sample 1000 times during a cycle. There would a thousand columns in Figure 20.5 instead of only ten. This would allow rapid changes in the analogue signal to be more accurately followed. The other way of improving the result is to have more binary digits (bits) in the conversion. In Figure 20.4 there are only four bits; this allows only 16 possible amplitude values in Figure 20.5. Given an eight-bit converter we obtain values ranging from 0000 0000 to 1111 1111 (0 to 255 decimal) making it possible to have 256 different amplitudes. With 16 bits, we can obtain 65 536, which is enough to reproduce the original waveform with extremely high precision.

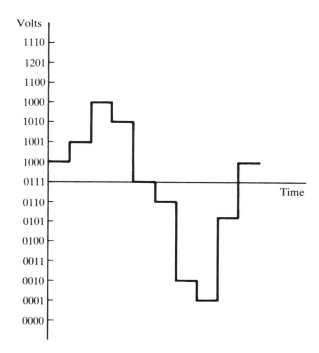

Figure 20.5 *The waveform of the digital signal shown in Figure 20.4*

It has been shown that to accurately reproduce the original waveform, sampling should take place at double the frequency of the highest frequency present in the sound. The highest frequency that is discernible by the human ear is about 20 kHz, so this is the highest frequency that we wish to reproduce faithfully. Accordingly the analogue signal should be sampled 40 thousand times per second. This is a high rate of sampling but well within the capabilities of modern logic circuits. The number of bits required depends on the application. An 8-bit ADC gives reasonable results with speech signals but for high-quality music recording and playback 16-bit conversion is preferred.

One of the prime advantages of a digital system is that, once a signal has been converted, it becomes a series of 0s and 1s, with no intermediate levels. In terms of tape recording, a zero is represented by a region of domains all magnetized in one direction. A '1' is represented by domains magnetized in the opposite direction. The tape is saturated in either direction. There are absolutely no problems associated with linearity, and no a.c. bias is required. Tape heads may be differently constructed and it becomes possible to record more data on a given length of tape. On reproduction, tape hiss due to non-uniformity of the magnetic particles and unevenness of the coating has no effect on the saturation of the tape, so is not heard on playback. When a digital tape is copied, 0s and 1s are copied as 0s and 1s; they do not change from one to the other. So a copy is an exact replica of the original. The effects of moderate tape and head wear have no effect on reproduction of digital signals. Thus many of the defects inherent in analogue tape recording are completely eliminated in digital recording.

There is another way in which digital techniques bring advantages. We are dealing with digitally represented numbers which can be handled at high speeds by methods like those used in a computer. They can be processed in various ways. When recording, there is no need to record the current sample on to the tape immediately. A whole batch of consecutive samples can be taken and stored in memory. Various operations can be performed on them before they are recorded (as a batch) on tape. One of the processes is to take the original numbers and code them in another form to make them more suitable for recording. We can add so-called check bits to the numbers so that, if an error occurs at a later stage, it can instantly be detected. Sources of error include blemishes and drop-outs in the tape coating which can cause a temporary break or distortion of the signal at playback. During playback the numbers recovered from the tape are decoded and analysed in batches, then sent to the DAC. If errors are detected, action can be taken to correct them, or at least minimize their effects. For example, if one sample is completely missing or (perhaps because on an error in one of its more significant digits) it is obviously too high or two low, the logic circuit can cancel that sample and substitute the averaged value of the two samples on either side of it.

Information recorded on the tape need not be limited to the sampled sound. Other data can be added as the recording proceeds. These include various codes to identify the batch number, the track number, number of tracks recorded, sampling frequency, and the time elapsed since the beginning of the recording. Information such as this enables the player to search for and play a selected track or tracks, or pick out a segment beginning and ending at a given time.

Digital audio tape (DAT) is recorded under a number of formats, the most commonly used one being R-DAT 1, which has become the standard. The tape is contained in a cassette of size similar to that of the compact cassette used in analogue tape machines. It is necessary to use tape based on iron or on barium ferrite. The tape is of the same width but is recorded and played on a pair of heads on either side of a rotating drum, in a way similar to that used in video recorders (p. 248). The tracks run diagonally across the tape (Figure 20.6), and hold blocks of signal data combined with blocks of control information. The samples of the two stereo channels are not simply placed side by side on adjacent tracks but are interleaved. Odd-numbered signals from the left channel alternate with even-numbered signals from the right channel on the same track. The even-numbered left channel signals and odd-numbered right channel signals appear on the adjacent track. This is made possible by using logical circuits to re-sort the blocks of data into a different order before transferring

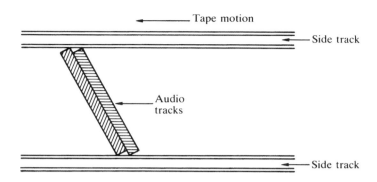

Figure 20.6 *The track layout on a digital audio tape*

them to the tape. This means that if there is a flaw in the tape this does not affect a large section of one particular track. The corrupted or missing data can be replaced automatically by data from the unaffected parts so that the damage is not noticeable on replay.

DAT has six modes of which three must be catered for in all recorders. The standard mode is that used for DAT recording and replay, with a sampling frequency of 48 kHz and 16-bit resolution. The mode for playing pre-recorded tapes has 16-bit resolution but the sampling frequency is 44.1 kHz. This is the same frequency as is used in compact discs (p. 212) and this allows manufacturers to produce both DAT tapes and CDs from the same masters. The third mode, for pre-recorded tapes, uses a wider track and faster tape speed to allow barium ferrite tape to be used. The other modes require the more expensive pure iron tape coating.

Digital compact cassettes

This is a more recent development of digital tape recording introduced by Philips. It is intended as a low-cost alternative to DAT and to become the more popular version. Costs are reduced by having a single stationary head with eight head-gaps in its top half. It lays down (or replays from) eight parallel tracks on the top half of the tape. At the end of play the head is turned over automatically and the direction of motion reversed so that the next half of the recording uses eight tracks on the bottom half of the track. The cassette is the same size as the ordinary compact cassette and players can accept and play ordinary analogue recordings. For analogue recording and playback, the head has two head-gaps on its lower half. The head is turned so that it uses the top half of the tape on the first run and the bottom half on the second (reversed) run.

The digital compact cassette uses a system of coding known as PASC (Precision Adaptive Sub-band Coding). This relies heavily on the abilities of specially designed logic ICs and is a good example of the complex processing that is possible with present-day logic circuits. The assumptions behind PASC are that the audibility of a low-level signal depends on its frequency. There are certain frequencies, particularly in the lower ranges that can not be heard unless their volume exceeds a certain level. It is a waste of resources to process, record and replay these, because they are inaudible. Also, a low-level sound made by one instrument is masked by a louder sound of the same frequency played by another instrument. It is a waste to record the low-level sound. To eliminate the unwanted sounds from the recording, the signal is passed through a number of digital filters (which have very sharp cut-off), dividing it into 32 segments each one third of an octave wide. The amount of signal present in each segment at any instant is assessed by the logic and 'empty' segments are not coded or recorded. Fewer bits are required for coding. In practice this technique yields the precision equivalent to 18-bit resolution, but needs only a quarter of the number of bits. Although this would appear to make high fidelity possible at low cost, there are sceptics who question the assumptions on which this technique depends.

Compact discs

Compact discs, or CDs as they are usually called, have largely replaced the older vinyl (LP) discs. One of the reasons for this is that CDs are digitally recorded and thus give better quality sound reproduction. There are incidental advantages such as their longer playing time (over an hour), their lack of wear, and their relative freedom from accidental damage.

The stages up to recording in the recording-reproduction chain are similar to those for DAT, except that different systems of coding are used and the additional information added to the data stream is appropriate to a disc system. Samples from the left and right stereo channels are recorded alternately. This is possible because of the digital nature of the process. Among the many processes of coding and error-checking the digital data before it is finally recorded, we can shuffle batches of data so that the two channels, although both are converted to digital simultaneously, can then be stored, sorted and put on to the disc in alternate batches. The operation is reversed at playback and both channels are heard simultaneously. The final digital signal is then recorded on a disc of optically flat glass coated with a photo-resistant substance. Recording employs a beam from a low-power laser focused on to the underside of the disc (Figure 20.7). Although the laser has a power rating of only a few milliwatts its beam is focused on to a spot only $0.1 \, \mu m$ in diameter. The intense energy of the beam at that point vaporizes the photoresist, leaving a pit about $0.6 \, \mu m$ in diameter and $0.1 \, \mu m$ deep. As the disc turns, the head produces a pit for

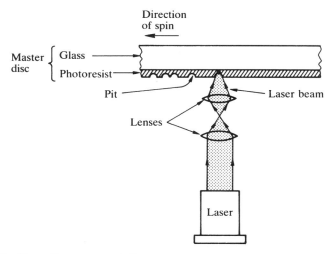

Figure 20.7   *Recording a compact disc*

every '1' in the digital signal, but leaves the disc unpitted when the digit is a '0'. The head is near the centre of the disc at the start of the recording and is moved toward the periphery as recording proceeds. Thus the pits are arranged along a spiral track with 41 250 turns. The rate of rotation of the disc is adjusted as the head moves outward, being gradually reduced to obtain a constant track speed of 1.2 m per second. Recording has to take place under scrupulously clean conditions as a single speck of dust could degrade several adjacent tracks.

A series of stages follow in which several copies of the master disc are made for use in mass-production. The discs eventually distributed in the shops consist of a transparent polycarbonate disc with dimples corresponding to the pits in the master. This disc is given a reflective coating (usually aluminium) which is then protected by having a polycarbonate layer sealed over it. Playback uses another laser, of much lower power than that used in recording. The optical system projects a finely focused laser beam on to the disc (Figure 20.8). The beam is partly refracted by the transparent plastic coating of the disc. Because of this the beam may be relatively broad where it enters the plastic layer; the plastic layer helping to concentrate it into a small spot on the reflective surface. Since the beam is broad where it enters the disc, the effects of small scratches on the surface are minimized. The beam strikes the reflective layer and, if no dimple is present, it is reflected back into the optical system. On its way back it is internally reflected in the prism, which directs the beam to a photodiode sensor. If a dimple is present, most of the beam is scattered sideways and little of it is reflected back to the diode. In this way the output from the diode consists of a series of low and high level representing the binary information stored on the disc. The signals from the photodiode are then sent to decoding

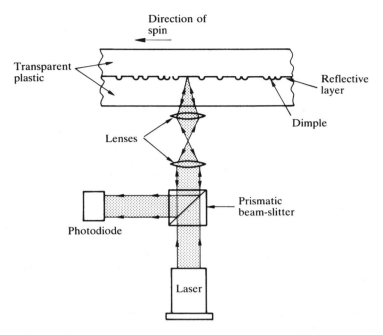

Figure 20.8 *Playing back a compact disc*

circuits which check it for errors, strip off the various information (such as track number and elapsed time) and decode it into a straight binary analogue of the original sound. This is then sent to the DAC which produces an analogue audio signal ready for amplifying and feeding to the loudspeaker.

Various methods are used to keep the playback head centred on the track. In one of these, the three-beam system, there is a tracking beam to left and right of the main beam, each with its photodiode. One beam is slightly ahead of the main beam and one is slightly behind. If the main beam is correctly centred, the two side beams skim the sides of the dimples equally and receive signals of equal intensity. But if the main beam wanders off track, one or other of the side beams receives a stronger signal. The timing of the signals from the two beams tell the logic whether the main beam is too far to the left or too far to the right. Whatever the error, the positioning mechanisms of the playback head is adjusted to correct it. Another system is used to ensure that the focusing of the beam is continually adjusted to give a spot of the minimum size.

CDs share with DAT the advantage that the recorded digital information remains unchanged no matter how many times the disc is played. There is no wearing of the grooves to degrade the reproduction as there is in a vinyl disc. But CDs are not indestructible and particular care must be taken not to scratch the outer surface of the plastic coating. The use of CDs as computer memory is described on p. 224.

Electronic music

Electronic musical instruments fall into two main categories: **keyboards** and **synthesizers**. A computer with a sound card and suitable software can function in the same way as a keyboard. When a musical instrument plays a note it emits a sine wave of given frequency, the **fundamental**. At the same time it produces a whole series of **harmonics**, which are sine waves of higher frequencies: multiples of the fundamental. The amplitude of the fundamental is usually greater than that of the harmonics and the higher harmonics usually have low amplitudes. The sound of a musical instrument is characterized by the harmonics that are present and their relative amplitude. The other feature that distinguishes the sound of a particular instrument is its **envelope**. Many keyboards are based on a number of oscillators running at the frequencies of notes of the musical scale. A selector switch is set to determine which instrument is to be simulated. Each time a key is pressed, logic circuits select the output from appropriate oscillators (for the fundamental and harmonics) and mix them (often using op amps) in their required strengths to make a sound resembling that of the instrument. Other circuits, using transconductance amplifiers impress the distinguishing envelope on the sound. The keyboard is also able to simulate percussion instruments, often by beginning with purely random waveforms (white noise) and filtering them and controlling their envelope to produce the required effects. An oscillator on the point of resonance may be used to simulate drums, gongs and guitar-like instruments. This is the method of sound production which is realistic for certain instruments but not for others. There is also scope for producing pleasing electronic sounds which do not attempt to imitate any other musical instrument. Note that this technique of sound production uses analogue circuits.

Frequency modulation synthesis is a simpler and more often used way of producing sounds. In this, the fundamental sine wave is modulated by a second sine wave of higher frequency. That is to say, its amplitude is made to vary according to the modulating wave. This produces a note rich in harmonics. By adjusting the frequency of the modulating wave and also the depth of modulation it is possible to create sounds resembling recognizable musical instruments, but more often the sound is interesting but is essentially 'electronic'. A more realistic way of simulating actual instruments is known as sampled-**wave synthesis or wave-table synthesis**. This is a digital method. The keyboard has a bank of memory in which are stored, in digital form, the recorded sounds of actual instruments being played. Only one note is stored for each instrument but when a note of given pitch is to be played it is processed by logical circuits to give it the correct frequency and to modify other characteristics of the sound.

Not only does the keyboard produce sounds but it has a battery of logic circuits to assist the performer to enhance the effects. With auto-chording, the player presses the keys singly but the instrument generates chords. The keyboard can also produce a rhythm accompaniment in a selected tempo, to back up the performance of the player. With these and other features the keyboard is a popular instrument, especially with the amateur player. A synthesizer is an instrument for the serious musician. It comprises banks of oscillators, mixers, delay circuits, reverberation circuits and others with which the player is able to create an immensely varied range of sounds.

21
Computers

From the supermarket checkout to the flight deck of a jumbo jet, computers are playing an increasing part in our lives and all this has been made possible only by developments in electronics. Computer technology is one of the biggest contributions of electronics to changing, and we hope improving, the world we live in. There are several kinds of computer, although the distinctions between them may be a trifle blurred. At one extreme are the **mainframe computers** operated by large companies and organizations. A mainframe computer usually occupies a room of its own and has a staff of operators. By contrast a **microcomputer**, also known as a **personal computer** or as a **desktop computer**, takes up very little room and is operated by a single person. In general, both types of computer can do the same things; it is just that the mainframe computer has the ability to handle larger amounts of data. But the emphasis of development has been on microcomputers and most of them now have greater computing power than the earlier mainframes. Instead of having a company mainframe computer there has been a trend towards providing staff members with individual microcomputers linked to each other on a local network. In the private sector there has been an immense increase in microcomputer ownership, and this tendency is set to continue. **Laptop computers** exploit the advances in electronic miniaturization and the development of large-area liquid crystal displays to provide a truly portable microcomputer.

The term 'computer' is not restricted to a machine with a keyboard and video monitor. Many other pieces of equipment incorporate computers to control their operations. These computers are dedicated to perform one particular task, such as controlling the operation of a clothes-washing machine. There are countless examples of such dedicated computers, found in satellites, guided missiles, CD players, Smartcards, and video games machines.

The main parts of a microcomputer appear in Figure 21.1. The functions of the various parts are as follows:

Clock

This is usually a crystal oscillator (p. 155) running at the rate of several megahertz (millions of cycles per second) to set the pace for the activities of all parts of the system.

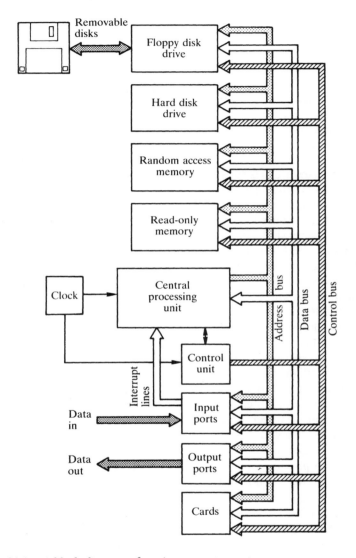

Figure 21.1   *A block diagram of a microcomputer system*

Control unit

Taking its cue from the clock, and its instructions from the central processing unit, the control unit is a set of logic circuits which coordinates the activities of the system.

Central processing unit

This is where calculations and logical operations on the data supplied to the computer, are performed by binary logic. The CPU consists essentially of a VLSI integrated circuit known as a **microprocessor**. This contains several sets of registers, each comprising a set of 8, 16 or 32 circuits of the flip-flop type (p. 150). These can be loaded with data in binary form. The data may consist of numbers, coded information, instructions to the microprocessor received from memory, or addresses of locations in memory. One of these registers, the **accumulator**, holds the data on which operations are currently being performed. In the accumulator the data may have other data added to or subtracted from it. The value held may be tested by comparing it with a given fixed value and subsequent action taken according to the results of the test. The data bits may be shifted along in the register in various ways. Individual bits may be tested to find out if they are '0' or '1' and appropriate action taken. These operations are all very simple in nature. The microprocessor does not in fact do anything very complicated. Its great power lies in its ability to perform many simple steps one after another at exceedingly high speed (millions of steps each second) without making mistakes.

Addressing

The memory of a computer consists of millions of sets of flip-flops or similar circuits. We need a way of locating one of these sets individually, either to read data from it or to store data in it. Like the houses in a street, each set is identified by its **address**. An address consists of 16 (usually, but possibly more) bits. With 16 bits, we can address up to 2^{16} ($= 65\,536$) different sets of flip-flops, but the memory of most computers is much larger than this. The addressing range can be increased by increasing the number of address lines or by working with separate blocks of memory which are switched on or off as required. Addresses are also allocated to input and output ports for use when receiving or transmitting data.

The CPU may sometimes have other ICs to supplement the action of the main microprocessor. There may, for example, be a **maths coprocessor** which is a microprocessor specialized for performing mathematical operations. This takes over the mathematical duties of the computer, leaving the main microprocessor free to handle input and output of data, storage of data, logical operations, and other non-mathematical tasks.

Input ports

These receive data from peripheral devices such as a keyboard, a mouse, or a modem. In industrial and scientific applications, input may come directly from electronic sensors or measuring circuits. An example is given on p. 208 showing how a varying voltage may be sampled and then converted to binary form by an analogue to digital converter (p. 200). The digital data is read in by the computer and displayed on the screen to resemble an oscilloscope display. There is more about input peripherals on p. 225.

Output ports

These transmit data to peripheral devices such as the video monitor and printer. Industrial computers may be connected to machinery (for example, a drilling machine) under their control.

Buses

These consist of sets of connecting lines (cables or tracks on a PCB) running in parallel. An address bus consists of 16 or more lines, each carrying one bit of a 16-bit (or longer) address. The advantage of having parallel lines is that all the bits of an address can be sent at once, which is much quicker than having to send the bits one after another along a single line. The address bus goes to all parts of the computer with which the CPU needs to communicate. When an address is put on to the bus by the CPU, it is received by all the devices connected to the bus. The address is decoded in each device (for example, in a memory IC) and, if the address corresponds to a memory location in that device, that location is put into contact with the data bus. The data bus, which may have 8, 16 or 32 lines, is used for sending data between the various parts of the computer. Data buses are bidirectional; they may be used by the CPU to send data to memory to be stored or to read data that is already stored in memory. The control bus links the control unit with other parts of the computer.

The data bus is communal; each part of the computer can take its turn (under the direction of the control unit) to send or receive messages along it, but only one device can be allowed to use the bus at any one instant. The control unit must allow only one device to put data on the bus at any one time. In this respect, the control unit is like the leader of a discussion group allowing only one person to talk at one time. But the situation is more complicated than this. Even when a device is not active, its outputs are at logical '0's and '1's. The outputs of all the inactive devices, if connected to the bus, would interfere with the output of the one active device currently using the bus. We must be able to disconnect inactive devices. To do this , we use a **tri-state output**. Figure 21.2 shows the output stage of a CMOS tri-state gate. The gate is controlled by an

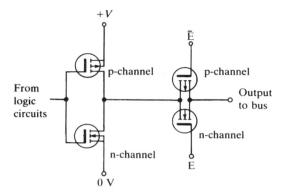

Figure 21.2 *Circuit of a CMOS tri-state output*

'enable' input, connected to one of the lines of the control bus. The input level (high or low) is passed through an INVERT gate in the IC, so that both the true (E) and inverted (Ē) states are available. If E is high, Ē is low, the output is enabled because both transistors of the complementary pair are conducting. Then the output of the gate is high or low, depending on the logic before it. It behaves just like an ordinary gate, sending data to the bus. When the device is disabled, E is low (and Ē is high). Both transistors are off. This completely disconnects the output terminal from the rest of the gate so it is unable to place data on the bus.

Interrupt lines

A signal, usually a low one on one of these lines causes the microprocessor to complete whatever operation it is engaged in and then jump to a special program known as an **interrupt servicing routine**. The interrupt signal comes from an external device, such as a modem which has just received a message over the telephone system. The modem has data that it has to send to the computer. In this case the interrupt servicing routine causes the microprocessor to accept the data and perhaps store it in memory. Then, depending on the situation, the computer might resume what it was doing before the interrupt. Interrupts are handled so rapidly that the user would probably not notice that anything unusual had happened. Then, at some more convenient time, or when asked to by the user, the computer looks for the stored message and displays it on the screen.

Read-only memory

The computer has a small portion of its memory, called ROM for short, which usually contains the program data that is needed as soon as the computer is switched on, to initialize the computer and make it ready to receive instructions from elsewhere. In specialized computers, the ROM may contain longer programs, such as word-processor programs, which are therefore ready to run as soon as the computer is switched on. It is not possible for the computer to store data in ROM, but it can read data that has already been stored there, usually by the computer manufacturer. When the address of a given memory location is fed to the address inputs of a ROM chip, the outputs go high or low, according to the data stored at that location. The CPU simply has to step through the addresses in order and a succession of instruction codes is put on the data bus to tell the CPU what to do next.

Data may be put into ROM by using special masks when making the memory chip, or, with a different type of ROM chip, it may be programmed into it at a later stage, before it is installed in the computer. Data written into the ROM during manufacture is permanent and is available every time the computer is switched on. Data written into a programmable ROM (or PROM, p. 195) is also permanent in the sense that the fusible links can not be re-formed once they have been blown. There are also electrically programmable ROMs in which data is stored by placing an electric charge on the gates of MOSFETs. If a gate is uncharged the output of the gate is low; if it is charged the output is high. Each gate is electrically insulated so that charges leak away extremely slowly. Data stored in such a ROM remains for years, so it is permanent for all practical purposes, unless there is reason to change it. There is a quartz window above the chip to allow the chip to be exposed to ultra-violet radiation, which removes the charges and destroys the stored data in a few minutes. This allows the ROM to be re-programmed, perhaps with an improved version of the program.

Random access memory

The CPU can both read data from this type of memory (RAM) and write data into it. Essentially it consists of number of flip-flops that can be set (output '1') or reset (output '0') by writing appropriate data into them. The flip-flops are arranged in groups of eight, each group having its own address, and providing storage for an eight-bit binary number or code, known as a byte. A typical microcomputer has several megabytes (million bytes) of RAM. The RAM is the workspace of the computer. It may be used to store the program that the computer is running, to store data on which it is to work, or to act as a 'scratch-pad' in which it can temporarily store the results of calculations. The essential point about RAM is that the CPU can store or read data at very high speed. This is essential if it is to be able to make full use of its ability to calculate and perform logic at high speed. However, there is a limit to the amount of RAM that a computer can conveniently use; 32 megabytes is the upper limit on many computers. This is partly due to the constraints of addressing a large memory and also the fact that having many RAM chips takes up space and requires extra power. This is why a computer needs additional storage, in the form of magnetic discs.

RAM differs from ROM in that the data stored in it is not permanent. All data in RAM is lost when the power is switched off. Any data that is to be retained must be transferred to magnetic or other storage devices before the computer is switched off.

Disk drives

Most computers have two kinds of disk drive. The **floppy disk drive** stores data on a thin flexible plastic disk coated on one or both sides with a magnetic film. Some computers are equipped with two such drives. Although the disk itself is floppy, and early ones were enclosed in flimsy cardboard covers, most disks are nowadays enclosed in a stiff plastic cover. The cover has a metal shutter which is slides back automatically when the disk is inserted in the drive to expose part of the disk surface to the magnetic head. The principle is the same as that used when recording music on digital audio tape (p. 207). The main difference is that the data is recorded on a number (often 40) concentric tracks (Figure 21.3), and the magnetic head moves radially to read or write each track. Each track is divided into sectors each one being allocated to one particular program or set of data. Longer programs or data tables may require more than one sector. There is a directory track on the disk telling the computer in which track and sector to look for each block of stored data and the magnetic head can skip from track to track and sector to sector, finding the information that is required. A typical floppy disk can store up to 1.4 megabytes of data. The disk spins at about 360 revolutions per minute but, even so, it still takes an appreciable time for the disk to be accelerated up to full speed, and for the magnetic head to be moved to the correct track and sector. Data can be read at rates of several hundred bits per second. A typical access time is 200 milliseconds, which is much slower than the 25 to 150 nanosecond access time of RAM or ROM.

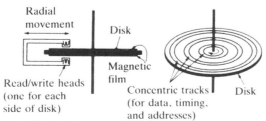

Figure 21.3   *The principle of the magnetic disk drive*

A **hard disk drive** has one or more disks attached to the same spindle. The disks are made of non-magnetic metal and coated on both sides with a magnetic film. The principle of storage is the same but the magnetic heads are much closer to the film. This is because the disks rotate at very high speeds (about 3600 revolutions per minute). This gives rise to a thin layer of moving air close to the disk surface in which the magnetic head 'floats' without actually coming into contact with the disk. Since the head is closer to the disk it is possible to record data more densely: the tracks are closer together and the recorded bits closer together than on a floppy disk. Consequently, a typical hard disk drive stores several tens or hundreds of megabytes and some have capacities as high as one gigabyte (a thousand million bytes). Another advantage of the hard disk is that the high rate of spin reduces access time to about 20 milliseconds. As the head is very close to the surface of the disk it is essential to exclude particles of dust or smoke. Hard drives are sealed during manufacture and can not normally be opened by the user.

Compact disc drives are very similar to CD players (p. 214) and work on the same principles. In fact some of them are able to play ordinary music CDs if the computer has a sound card (see below) installed. The information stored on a CD is simply a series of 0s and 1s. It may represent musical sounds but it could equally well be used for storing information of other kinds. In computing terms, a CD stores about 600 megabytes of data. This is more than is stored on a typical PC hard disk, and it would require over 400 3.5" double-density floppy disks to store the same amount of information. CD drives are used as memory storage devices for computers, the main difference being that they are read-only memory (CD-ROM). It is not possible to 'save' information from the computer onto a blank CD. CDs are at the centre of the new developments in **multimedia technology**. A disc can store text, computer programs, photographs and diagrams, motion pictures and sound. These can be accessed and loaded into the computer almost instantly. Very elaborate games with startling graphics are now available on CD-ROM but more serious applications of the technology include educational and reference discs. A specialist application of CD-ROM is the **Photo CD** (or PCD). This stores a number of photographic images in digital form. The information can be played out to show the photograph on a TV or on a computer screen. It can also be sent from one site to another by all the usual means of data transmission. The Kodak PCD is able to store up to 100 photographic images in full colour. The first stage in production is to take the photographs with a camera on ordinary colour transparency film. The transparencies are each projected by a high-quality slide projector on to an array of photosensitive electronic devices. The digital data so obtained is coded and then stored on the CD. The discs can store up to 640 megabytes of data. They have a gold reflective layer, instead of the usual silver, and this is cut away by the action of a fine laser beam (p. 213) during recording to produce a sequence of pits, with lands between them, representing the digital data. The data may be read from a PCD by an ordinary CD-ROM drive attached to a computer. It can also be read by a dedicated PCD player attached to a TV set or monitor. A PCD player is similar to an audio CD player and plays audio discs too.

Each photograph is stored five times in files offering different degrees of resolution. The lowest resolution gives a small picture on the screen, suitable for identifying the picture and showing its main features, in much the same way as we make contact prints of 35 mm transparencies for indexing a slide collection. The next highest resolution is suitable for TV or monitor display at low resolution, while the next above this gives excellent pictures on the TV or monitor screen. The two highest levels of resolution produce pictures suitable for high quality enlargements and for showing on a high-definition TV screen. Before the photographs are recorded it is possible to edit the images, cropping them, enlarging parts of them or altering the colour balance. The data can not be overwritten once it is recorded, but it is possible to record data on a partly-full disc in several sessions. This means that a collection of photographs can be built up on the disc as batches of photographs are taken and developed. As with audio CDs, PCDs are durable and, provided they are properly treated, give the same high quality every time they are played. Copy discs are just as good as the original.

Cards

Most computers have sockets for plugging in optional circuit-boards (known as cards) to extend the capabilities of the system. These may include **modem cards** for communicating with the telephone network, **fax cards** to allow the computer to send and receive faxes, and **sound cards** to add sound producing and recording facilities to the computer. Another type of card is the **memory card** which carries a number of RAM ICs, enabling several hundred kilobytes of data to be stored and read. Memory cards have a built-in lithium battery which powers the ICs whenever the card is removed from the computer, so that the data are not lost.

Input peripherals

Keyboard

Most computers have this peripheral as standard, usually in the shape of a typewriter QWERTY keyboard, with a numeric keypad on one end and various other function keys. Each plastic key-top is held up against the plastic frame by a spring, with guides to keep it in position. There is a double plastic membrane beneath the frame, one with parallel tracks corresponding to the rows of keys, the other with slanting parallel tracks corresponding to the columns of keys. When a key is pressed, a track in one layer is brought into contact with a track in the other. The computer is programmed to scan the rows and columns repeatedly whenever it is waiting for input. If it detects a contact between a given row and a given column, it knows which key has been pressed. On p. 198 we mentioned contact bounce, which is always a problem with fast-acting logic circuits. Debouncing each of the hundred or more keys of a computer keyboard by the method shown in Figure 19.8 is not practicable, but a 'software solution' is given on p. 230.

Mouse

Users who are not familiar with a keyboard may find a mouse a faster and more convenient way of telling a computer what to do. A mouse consists of a plastic box shaped so as to fit conveniently under the hand. The 'tail' of the mouse is the lead by which it is connected to the computer. A plastic ball projects slightly from the under-side of the mouse and rests on the surface on which the mouse is placed – usually a plastic 'mouse pad'. When the mouse is moved, the ball rotates. There are three plastic wheels in contact with the ball, two of which rotate around perpendicular axes to detect left-right and up-down motion on the mouse pad. On the same spindle as each wheel is a sectored disc. Light from an LED passes between the sectors and is detected by phototransistors. The pulsed signal produced as the discs rotate indicates the rate of motion of the mouse in the two directions. This can be interpreted by software and used to position the cursor on the monitor screen. The mouse has up to three buttons (only one is generally used) which close microswitches. These are used to signal instructions to the computer, for instance to indicate that the cursor is now on a part of the screen selected by the user. When the button is pressed ('clicked') the computer performs an appropriate action.

Trackball

This has the same function as a mouse but consists of a ball rotated by the fingers in a socket at the front of the keyboard. It does not require as much desk space as a mouse mat, so it is ideal for use with laptop computers.

Digitizing pad

In one type of pad, a grid of wires is embedded in a rectangular plastic tablet (Figure 21.4). The computer sends a pulse to each horizontal wire and each vertical wire in turn, thus scanning the whole pad at high speed. The pulses are detected by a sensor (such as a Hall-effect device) at the end of a stylus and a pulse is sent to the computer. By this means the computer can determine the exact location of the stylus tip on the pad. It can be programmed to display a dot at a corresponding location on the computer screen. As the stylus is moved

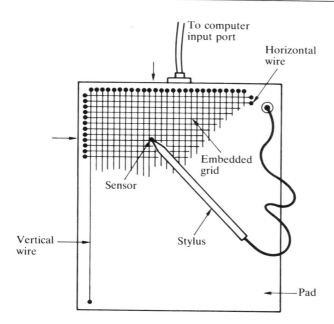

Figure 21.4 *A digitizing pad. The sensor in the stylus sends a signal to the computer when a pulse passes along the two wires indicated by the arrows*

a line is drawn on the screen. Digitizing pads are used for entering graphical data. Rough drawings can be tidied by software and processed into a form suitable as artwork for publication.

Touch screen

The monitor screen has a shallow frame in front of it in which are located an array of ultrasonic or infra-red sources and sensors. When a person touches the screen, placing the finger on a displayed panel or 'button', signals from the sensors indicate the finger position. Depending on the position of the finger, the computer takes appropriate action. Another method uses a screen coated with transparent conductive strips. The capacitance between the finger and one of the strips indicates where the finger is pointing. In general, touch screens are limited in their usefulness by their low resolution.

Scanner

A document such as a photograph or a typewritten sheet is scanned by moving a spot of light over it. Variations in the amount (and in a colour scanner, the colour) of the light reflected from the document are read by the computer and used to build up an image of the document on the screen. Software is available that will process the scanned image of printed material and recognize the characters. It is then able to convert this to a word-processor document with remarkable accuracy. Scanning a document into a computer and correcting a few errors is much quicker than copying it by typing on the keyboard.

Magnetic stripe readers

The resistance of ferromagnetic materials depends upon the strength of the local magnetic field. In a **magneto-resistive sensor**, a current from a constant-current generator passes along a strip of ferromagnetic material. As the local field alters, so does the resistance of the strip and therefore the p.d. across it. The variations of p.d. can be measured. Inductive magnetic sensors depend upon the rate of change of magnetic field, but magneto-resistive devices depend solely on field strength. These sensors are used for reading data recorded magnetically on tapes and computer disks. They are also used for reading from the magnetic stripes on bank cards, tickets, and identity cards. Because they do not depend on the rate of change of field strength, they are unaffected by the rate at which the magnetic stripes are swiped through the reading machine.

Optical disks

Although computer data is most often stored on magnetic disks, other forms of data storage may sometimes have advantages. One of these is the optical disk, in an optical disk drive. An optical disk consists of an active layer sandwiched between two transparent layers. Data is recorded on the disk by directing a high-powered laser beam on to its surface. The beam heats and melts a small spot on the disk to correspond with a '1'. With one type of active layer, the material crystallizes as it cools. This changes the reflectivity of the surface, producing a series of spots that are more reflective than the unmelted material. The disc is read by directing a low-power laser beam on to it and detecting the amount of light reflected. Such a disk is a 'write once read many' or WORM disk. Another type of active material can be used which is already crystallized and has high reflectivity. A high-power beam is used for writing, destroying the crystalline structure and so producing spots with lower reflectivity. This can be read using a low-power laser beam, as above. But if the writing beam has medium power, this is sufficient to restore the crystalline structure and can be used to erase data already written on the disk. As the writing laser scans the disk, it can be energized at high power where it is to produce a new spot or at medium power where it is to erase a previously recorded spot. Thus the disk can be overwritten with new data.

Magneto-optical disks

Magneto-optical disks in a suitable drive are another technique used for storing computer data. They store data as a pattern of bits by directing a laser pulse at the magneto-optical material in a magnetic field. The laser beam melts the material, the molecules of which orientate themselves in the field. The material immediately cools, leaving the small spot permanently magnetized, representing a logical '1'. The disk is read by using a polarized laser beam. Magnetization alters the reflective properties of the surface of the disk. As a result of this, the direction of polarization of the beam is changed if it is reflected from a magnetized spot. This change is detected and used to distinguish between magnetized ($=1$) and non-magnetized ($=0$) regions. Such a disk can be written to repeatedly to alter the stored data.

One limitation on the amount of data that can be stored is the size of the spot that can be heated, in other words the diameter of the laser beam. Normally lasers emit radiation in the red region of the spectrum, where wavelengths are longer, and the minimum beam diameter is 0.8 µm. Increased data density is obtained by using a blue laser which can be focused to a spot only 0.4 µm in diameter. This has only one quarter the area of the beam from a red laser, allowing data density to be increased up to four times. A magneto-optical disk the same diameter as a CD can store up to 6.5 gigabytes when written to by a blue laser. This is enough room for 6500 novels each 250 pages long. Data may also be stored on miniature magneto-optical disks that are only 2.5 inches in diameter and which hold up to 140 megabytes.

Hardware and software

We have already mentioned software in passing but now we look at the concept of software more closely. When talking about a computer system, engineers divide it into two general parts: hardware and software.

Hardware refers to the physical units such as the central processing unit, the memory, or the drives that make up the computer.

Software, on the other hand, is the term used for the programs that control the operation of the computer through the CPU and enable calculations, logical operations, fault detection and other functions to be performed automatically. Software is usually stored on magnetic disks, mostly on the hard disk, which has ample room for long programs. When a program is to be run, the software is copied from the disk into RAM. Some programs are so long that there is not room for the whole program to be held in RAM at any one time. In such cases, different parts of the program are transferred to RAM from the disk as and when they are required. Conversely, if RAM holds a lot of stored data that is not currently required, it can be transferred to the disk and re-loaded back into RAM later. All of this swapping between RAM and the disks is performed automatically according to the operating system of the computer.

Very often it is possible for software to take over the functions of hardware. An example of this is the debouncing of keys on the keyboard. This could be done by having a capacitor and inverting gate connected to each key (Figure 19.8) but software offers a much simpler solution. When the computer detects a key press, it registers the fact but does nothing more immediately. At intervals of a fraction of a second it repeatedly scans the keyboard, registering the key-press each time. After it has found the key to be pressed on several consecutive scans it accepts the fact that the key is pressed and takes appropriate action. This routine can also be used for avoiding the effects of rollover, when a key is pressed before the previous one is released so both are held down at the same time. The computer can decide which key was pressed first and which next.

22
Telecommunications

The term 'telecommunications' means 'communicating at a distance.' This could include communicating by sending a letter by post but the use of the term is restricted to electrical or electronic communications. Information may be transmitted either by electric currents (in wires) or by electromagnetic radiation. Electromagnetic radiation is usually taken to mean radio waves but visible light and infra-red are also forms of electromagnetic radiation, so communication by optical fibre is included in teiecommunication. The term includes the transmission and reception of sound, vision, text and information of other kinds. Formerly all telecommunication was analogue in nature (except perhaps telegraphic transmissions in the Morse Code, which can be thought of as digital) but there is an ever-increasing trend toward digital transmission. Developments in electronics have made possible the worldwide boom in tele-communications, enabling information to be sent instantly to all parts of a country and to all parts of the world. Electronics has made telecommunications an important factor in all our lives.

Telegraphy

This is the most primitive form of telecommunications, requiring only a pair of wires, two switches (keys), two lamps or buzzers and a power supply. It was formerly used to send messages over distances of a few kilometres. The Morse code, with its system of dots and dashes to represent letters and numerals, was adopted as the internationally recognized system for transmitting information. It has been almost completely superseded by voice communication (telephony), although the basic idea of telegraphy still persists in the domestic door-alert and similar systems.

Telephony

Originally this was a wholly analogue system. The telephone hand-set incorporates a microphone and a loudspeaker. Analogue signals are sent along a pair of wires to the local exchange. All the instruments in the area are connected to this exchange and here it is possible for connections between any two pairs of instruments to be made. The original manual switchboard was first replaced by a system of mechanical switches which were controlled remotely by pulses 'dialled' by the person making the call. Dialling '5', for example sent 5 pulses to the exchange and the mechanical switch was stepped on 5 steps to position 5. Nowadays the switching is done by logical means, using digital circuitry. The telephone instrument includes an oscillator circuit capable of producing a range of tones. The oscillator is an IC driven by a 1 MHz clock and which generates digital pulses at eight different frequencies (tones). The IC is controlled by a keyboard on the telephone and whenever a number key is pressed it produces two of these tones. For example, pressing key '5' produces tones at 770 Hz and 1477 Hz. The dual signal is sent to the exchange, where there is a circuit to detect which frequencies are present. In this example the circuit would identify the two tones as 770 Hz and 1477 Hz and register the fact that a '5' has been dialled. Generating and recognizing tones takes a fraction of a second and the response of the electronic switching circuits is virtually instantaneous, so that connecting a call is very much quicker than it used to be.

The principle of linking all telephones in an area to a central exchange is extended to linking all exchanges in a country or part of a country to major exchanges by long-distance line. Ultimately a world-wide network of exchanges is built up.

High frequencies

Frequency of oscillation is measured in hertz, symbol Hz. A frequency of 1 Hz is one oscillation per second. The frequency of the a.c. mains is 50 Hz, sometimes referred to as '50 cycles per second'.

For high frequencies we use the familiar prefixes:

$1000 \, \text{Hz} = 1$ kilohertz (kHz)
$1000 \, \text{kHz} = 1$ megahertz (MHz)
$1000 \, \text{MHz} = 1$ gigahertz (GHz)

Although a wired telephone system may seem relatively easy to operate, there are problems to be overcome. One is that the resistance of telephone wires depends on their length, so the power of the signal falls off with distance. For long-distance communication it is necessary to install amplifiers, or repeaters, at intervals along the line. There is also the problem of line impedance. When a pair of conductors are laid side-by-side there is capacitance between them. In addition the wires have self-inductance. This may not be apparent when the wires are short but with telephone lines the effects of resistance, capacitance and inductance amount to an appreciable line impedance, which has to be taken into account. The mathematics of line transmission is too complicated to be dealt with here but one of its consequences is easily understood. It was stated on p. 170 that for maximum transfer of power between one part of a circuit and another the output impedance of one part must be equal to the input impedance of the other. The same applies to the parts of a telephone circuit, including the impedances of the connecting lines themselves. Impedances must be matched if maximum transfer is to be achieved. This is essential to ensure that the signal reaches its destination loud enough to be heard, but this is not the only reason. If maximum power is not transferred from one part to another, some of the power must go elsewhere. If there is an impedance mismatch, part of the signal may be reflected back along the line, causing an echo. This causes serious degradation of analogue voice signals and can wreak havoc with digital signals.

With the advent of digital circuitry the telephone lines have been put into use for other forms of data transmission. Communication, particularly long-distance communication, between exchanges is often purely digital and may involve fibre-optics (see below). By using digital methods it is possible to compact the data and to transmit several sets of data at once along the same lines, and at high speed. Only in this way is it possible to maintain today's vast worldwide traffic of information, culminating in the so-called Information Superhighway. Nowadays it is easy for a person sitting at a computer in an office in, say, New Zealand to communicate with a computer in, say, Edinburgh to access instantly data stored in that computer's memory and to run software held in that computer. The time has arrived when anyone with a computer and telephone modem is able to access information from anywhere in the World.

Although national and international data transmission is becoming increasingly digital, connections to the local exchange still make use of the existing lines intended for analogue signals. The technique for transmitting digital data along these lines is usually that known as frequency shift keying (or FSK). This is the system adopted when a computer uses a modem to send and receive data. The computer produces data in digital form and usually sends it byte by byte to the modem along eight parallel lines. The modem has an IC known as a universal asynchronous receiver/transmitter (or UART) which converts the data from parallel to serial form. That is to say, it transmits the individual bits of a byte along a single line, one after another. It also adds several extra bits, such as a start bit, to indicate the beginning of a sequence of bits from the byte, a parity bit, and one or two stop bits to indicate the end of the sequence.

Parity

This is a way of detecting errors in data transmissions. The parity bit may be '0' or '1', and its value is selected automatically so as to make the total number of '1's sent an even number. On reception, the number of '1's received is checked and, if it is found to be odd, an error must have occurred. This system is called **even parity**. If preferred, a UART can operate under odd parity in which the parity bit makes the number of '1's odd.

The sequence of '0's and '1's is sent to the telephone line by the modem IC as a sequence of two tones. There are various systems, but typically a '0' is sent as a burst of tone at 1070 Hz. A '1' is sent as a 1270 Hz tone-burst. These two tones are generated in the modem from a single 1170 Hz tone, which is shifted up or down by 100 Hz to obtain the two tones required. This is why this technique is called **frequency shifting**. At the receiving end, the frequencies are detected by the receiving modem, which produces the corresponding logical '0's and '1's. A UART takes this stream of bits and re-converts it to bytes, removing the stop and start bits and checking the parity bit as it does so. The bytes then go to the receiving computer. Simultaneous two-way communication between the two modems along the same telephone lines is accomplished by using a second pair of tones. The modem originating the call uses 1070 Hz and 1270 Hz, as described above, and the answering modem uses 2025 Hz and 2225 Hz.

Facsimile

Perhaps the most significant addition to telecommunications equipment has been the **fax machine**. Fax machines are used to transmit exact copies (or facsimiles) of documents over the telephone lines. The documents may be hand-written if preferred and it may include drawings and photographs, which greatly adds to the usefulness of this method of communication. A fax machine is really a microcomputer dedicated to perform certain tasks automatically. It includes a dialling pad and circuitry, and usually a memory for telephone numbers for one-touch dialling of frequently called numbers. It has a scanner which moves the document past a photo-sensitive line scanner. This produces a sequence of digital signals which is put on to the telephone line by a built-in modem. The receiving machine prints out the document using a thermal printer.

If a computer has a fax card installed and is connected through a modem to the telephone network, faxes may be sent and received directly by a computer. The fax is composed just like any other computer document, using a word-processor or graphics program. Sending the document to the fax card or modem is almost the same as sending it to a printer. Received faxes may be viewed immediately on the monitor screen. They may be saved to disk or printed out.

As telephone charges continue to fall because of improved technology and postal charges continue to rise, faxes become relatively cheaper. Already it can cost less to send a single-page fax than to send a single-page letter by post, and transmission of faxes is virtually instantaneous.

Optical fibre

In Figure 22.1 a beam of light is passing through a transparent material (such as glass). When it reaches the boundary between the glass and the surrounding air, one of two things may happen to it. If the angle at which it meets the boundary is sufficiently small (as measured between the beam and a line perpendicular to the surface), the beam passes out of the glass into the air; it is bent at the point at which it emerges. But if the angle exceeds a certain amount (the **critical angle**) the beam is reflected. This is called **total internal reflection** because none of the beam passes out of the glass. Given a narrow fibre of glass, as in Figure 22.2, a beam directed into the end of the fibre is reflected repeatedly until it finally emerges from the other end of the fibre. Provided the glass is of high quality and maximum transparency, very little light is lost. As shown in Figure 22.2, the fibre does not have to be perfectly straight. Curves in the fibre make no difference to the transmission of light.

In a practical fibre-optic cable the glass fibre is about 0.1 mm in diameter. Usually it is surrounded by cladding, consisting of a layer of glass or plastic; the optical properties of the cladding are such that total internal reflection occurs at the boundary between the fibre and the cladding. The cladding helps protect the fibre from surface scratches which would lead to loss of light. Several such jacketed fibres can be surrounded by a plastic sheath to form a multi-core cable only a few millimetres in diameter, each fibre being

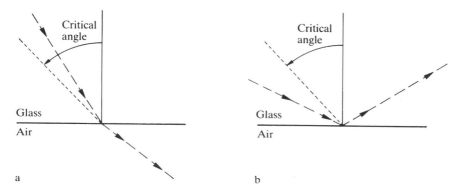

Figure 22.1 *The ways a ray of light travelling in glass behaves when it strikes a glass-air surface (a) Refracted (b) Totally internally reflected*

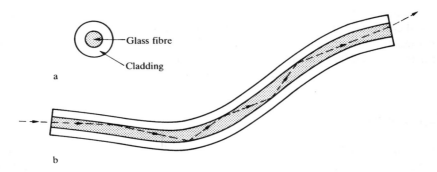

Figure 22.2   *Optical fibre (a) as seen in cross-section (b) as seen in lengthways section showing the light being totally internally reflected*

able to carry a separate signal. At the transmitting end of the fibre is an LED or laser diode which produces the light. At the other end is a photodiode. If the link is a long one, there are repeaters to detect a weak signal and relay it as a strong one. Since signalling is digital there is no loss of quality on repetition.

In practice, each fibre carries several signals simultaneously, each having its own frequency. It is like having several radio stations broadcasting simultaneously at different frequencies (see later). At the receiving end each signal can be separated out (in a similar way to that in which we tune a radio receiver to just one station) and processed independently. Because of the high frequency of light, it is possible to carry many more simultaneous transmissions than can be carried on copper-wire cables. This is one of the main reasons for the adoption of fibre-optics for telecommunications. Another reason is that the cables are light in weight and inexpensive compared with copper cables. Unlike copper cables, they are unaffected by electromagnetic interference and consequently they are more secure. It is possible to use surveillance equipment to detect and read signals passing through a copper cable, but this is not feasible with optical fibre. For these reasons, optical fibre has been extensively installed on the main national and international trunk routes of the telephone system.

Radio

Radio or wireless transmission is the alternative to telecommunication by wire or optical fibre. Radio is a form of electromagnetic radiation, and like all electromagnetic radiation, is produced when atomic particles oscillate at high frequencies. It is the acceleration of the electrons which produces the radiation, which means that radio waves are generated by a lightning strike or when we produce a spark by opening a switch. Some of the earliest transmitters relied on the production of sparks to generate the signals. In modern radio transistors, we accelerate the electrons to move to and fro in a conductor, the aerial or antenna.

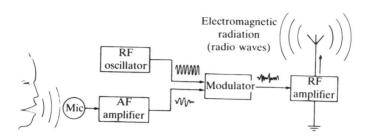

Figure 22.3   *Block diagram of a radio transmitter*

Figure 22.3 shows the main parts of a radio transmitter. The radio frequency oscillator is an oscillator circuit designed to produce sine waves at high frequency, most often in the range 300 kHz to 3 GHz. Like the circuit of Figure 15.8 the frequency of oscillation is determined by the values of the capacitor and inductor in a resonant circuit. Alternatively a crystal oscillator (p. 155) is used. Given the frequency, we calculate the wavelength of the electromagnetic radiation from this equation:

$$\text{Wavelength} = \frac{300\,000\,000}{\text{Frequency}}$$

where the wavelength is in metres and the frequency is in hertz. The number 300 000 000 is the approximate velocity of light (and other electromagnetic radiation, including radio waves) in metres per second.

Frequency and wavelength

Description	Frequency	Wavelength	Application
Audio frequency*	30 Hz to 20 kHz	11 m to 17 mm	Audible sound
Very low frequency radio	30 Hz to 30 kHz	10 000 km to 10 km	Experimental
Low frequency radio	30 kHz to 300 kHz	10 km to 1 km	Telecommunications
Medium frequency radio	300 kHz to 3 MHz	1000 m to 100 m	Telemetry, control telecommunications
High frequency radio	3 MHz to 30 MHz	100 m to 10 m	Telecommunications
Very high frequency radio	30 MHz to 300 MHz	10 m to 1 m	Telecommunications and control
Ultra-high frequency radio	300 MHz to 3 GHz	1000 mm to 100 mm	Television, telecommunications, navigation
Super-high frequency radio	3 GHz to 30 GHz	100 mm to 10 mm	Radar, microwave, telecommunications and other devices

* Vibrations in air

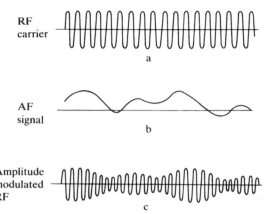

Figure 22.4   *Modulating the amplitude of a carrier wave (AM)*

The output of the oscillator may be fed directly to the aerial and a signal sent by switching the oscillator on and off. By this means we could send a message by transmitting the dots and dashes of the Morse code. More usually the radio-frequency oscillator is connected to a circuit known as a **modulator**. Instead of turning the oscillator on and off, we keep it running all the time but modulate its output. One way of modulating the signal is shown in Figure 22.4a, b and c. An audio signal is detected by a microphone (Figure 22.3) and amplified. The amplified signal is sent to the modulator which uses it to vary the amplitude of the radio frequency signal. In Figure 22.4a we show the unmodulated radio-frequency wave and in Figure 22.4b we show the audio signal. Its frequency varies with the pitch of the voice but is always much less than that of the radio frequency signal. In Figure 22.4c we see the results of modulation; the signal has its original radio frequency but its amplitude varies according to the audio signal. This is referred to as **amplitude modulation**, or **AM**.

Another method of modulation is **frequency modulation**, or **FM** (Figure 22.5). Which is rapidly replacing AM as the method used for national and regional broadcasting. This uses a different kind of modulator circuit which, instead of modulating the amplitude of the signal, modulates its frequency. The frequency of the modulated signal retains its average value produced by the RF oscillator but is sometimes slightly higher and sometimes slightly lower, depending on the audio signal. FM works well only when the radio frequency is high.

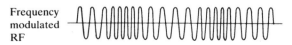

Figure 22.5   *Modulating the frequency of a carrier wave (FM)*

In both systems of modulation the unmodulated wave is known as the **carrier wave**, since it carries the audio signal in its modulation. Simply turning the transmitter on and off to produce Morse code signals as described earlier is known as an **interrupted carrier wave**.

The form of the transmitter aerial is mainly dependent on the frequency being transmitted. For the lowest frequencies, the aerial may be a length of wire suspended above the ground on insulated supports (Figure 22.6). Such a **long wire** aerial is connected to the output of the modulator. The 0 V or ground line of the transmitter is earthed by wiring it to a metal spike embedded in the soil. Electromagnetic radiation spreads in all directions from the long wire aerial. Much of it passes upward into the higher regions of the Earth's atmosphere. An active layer in the atmosphere, known as the ionosphere reflects the radiation several times and eventually it returns to the surface many hundreds or thousands kilometres from the transmitter. For this reason, low-frequency and medium-frequency transmission is best for direct long-distance telecommunications, as opposed to indirect communication, using satellites (see p. 259). It is however subject to fading if the ionosphere becomes disturbed or weakened, often as the result of variations in sunspot activity.

For VHF and UHF transmission the most favoured aerial is a **dipole**, similar to the receiving dipole illustrated in Figure 22.8a. It is connected to the output and ground line of the modulator. The propagation of the waves is strongest in the directions perpendicular to the length of the dipole elements. If the aerial is vertical, the radiation is strongest in the horizontal direction. For good reception, the receiver should be in sight of the transmitter, which is usually situated on high ground to cover maximum area. Thus VHF and UHF stations are reliably received only locally and a network of transmitters is required to service a whole country.

Satellite telecommunications and the transmission and reception of super high frequency radio (microwaves) are described in the next chapter.

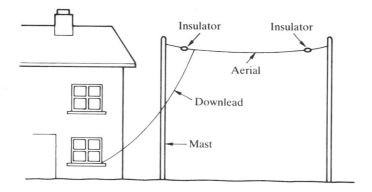

Figure 22.6 *A long-wire aerial*

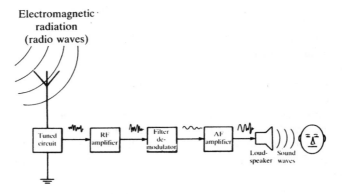

Figure 22.7   *Block diagram of a radio receiver*

Radio reception

The essential elements of a radio receiver are shown in Figure 22.7. The aerial may be a long wire (Figure 22.6) or a dipole (Figure 22.8a) , depending upon the frequency to which the receiver is tuned. Currents produced in the aerial on the arrival of a radio signal are fed to a tunable resonant circuit consisting of an inductor and variable capacitor. A dipole may have additional elements to improve reception in fringe areas (Figure 22.8b). Only the dipole is connected to the receiver the other two elements being unconnected. The lengths and spacing of the elements are related to the wavelength. Signals of the correct wavelength cause currents to flow in the elements which re-radiate electromagnetic waves. The result is that signals reaching the aerial from the direction of the director are received more strongly than those from other directions.

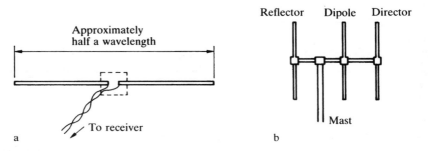

Figure 22.8   *Dipole aerials (a) Simple dipole (b) A three-element dipole*

Domestic receivers in the low- and medium-frequency bands often have an internal aerial in the form of a ferrite rod. Ferrite has a high magnetic permeability compared with air so the magnetic lines of force of the arriving radio signal tend to be bunched together in the rod instead of passing on either side of it. This produces an alternating field in the rod. One or more coils are wound around the rod and alternating currents are induced in these. The ferrite rod with coils around it is connected in parallel with the tuning capacitor of the radio receiver, forming a tunable resonant circuit.

Whether the aerial is a long wire, a dipole or a coil on a ferrite rod, the result is the same. When the resonant circuit is tuned to the frequency of the carrier wave it resonates strongly in sympathy with the carrier. The amplitude of the oscillations varies according to the amplitude of the carrier which, as we have seen, is an analogue of the audio signal. This signal is amplified by a radio-frequency amplifier. It then goes to a rectifier circuit or detector circuit which, in a simple radio receiver may consist of nothing more than a diode. The action of the detector is seen in Figure 22.9a and b.

The next stage is a low-pass filter, which removes the high-frequency (that is to say, the radio-frequency) component from the signal. We say that the signal is **demodulated**. This leaves only the audio-frequency component, equivalent to the average voltage shown by the dashed line in Figure 22.9b. This is fed to an

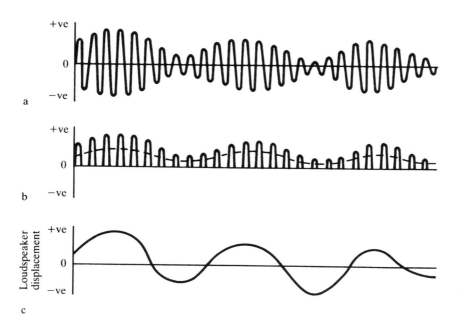

Figure 22.9 *Detection and amplification of a radio signal (a) The signal as received in the tuned circuit (b) Rectified signal, with the average voltage shown by the dashed line (c) The audio signal as sent to the loudspeaker*

audio-frequency amplifier which increases its amplitude and makes it alternate about zero volts. The amplified signal is sent to a loudspeaker and the original broadcast sound is heard. In an FM receiver demodulation consists in comparing the incoming signal with a signal generated in the set and having the carrier frequency. The demodulator produces a signal dependent upon the difference between the frequency of the received signal and the carrier frequency at each instant. This difference is the audio signal. An advantage of FM is that the demodulator is not affected by variations in the amplitude of the received signal. Variations of strength due to fading have only a limited effect. Sudden bursts of radio noise such as caused by lightning are very noticeable on AM, but have no effect on FM.

Selectivity

One of the problems of radio reception is that of picking out the transmission from one particular station from a host of other transmissions. Using a directional aerial such as a dipole with one or more director elements helps to pick out the required station and a ferrite aerial too is directional. To pick out a signal station we have to rely on the selectivity of the resonant circuit. Although this in theory resonates most strongly at the frequency to which it is tuned, the circuit is able to resonate quite strongly at close frequencies. For this reason a strong local station with a frequency close to the tuned one may swamp reception of the required station. It is possible to design a circuit to make it more selective and respond more strongly to one particular frequency but, unfortunately, the more selective the tuning circuit is made, the less sensitive it becomes. Distant and weak stations are not received.

The **superheterodyne receiver** (or superhet) is a solution to this problem in AM receivers. The superhet depends for its action on the phenomenon of **beats**.

Beats

When two signals which are almost equal in frequency and amplitude are mixed together, the result is a signal which changes regularly in amplitude, causing beats. With two sound signals we hear a throbbing sound, the rate of throbbing or beating being equal to the difference between the two frequencies. In Figure 22.10, we show two sine waves with frequencies 200 Hz and 220 Hz. When added together, the resulting signal (thicker line) varies in amplitude, or beats, at 20 Hz.

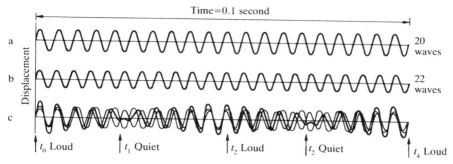

Figure 22.10 *Interference between two waveforms produces beats (a) Wave at 200 Hz (b) Wave at 220 Hz (c) Combined wave shows beats at 20 Hz*

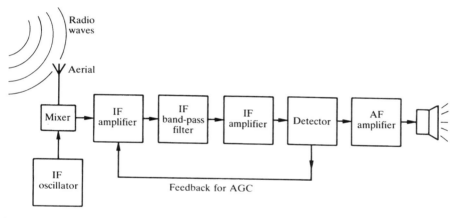

Figure 22.11 *Block diagram of a superheterodyne radio receiver*

It has its own oscillator (Figure 22.11) operating at an intermediate frequency (IF). When the set is tuned to receive a station at a given frequency, the internal oscillator is tuned to a frequency that is (usually) 455 kHz above the station frequency. The two signals are mixed and this produces a new radio-frequency signal which beats at a frequency equal to the difference between the station frequency and the intermediate frequency. As we have already said, this difference is 455 kHz so, whatever the frequency of the station the signal passed to the rest of the receiver is at 455 kHz. This is the new carrier frequency and the audio signal is still amplitude-modulated onto it, just as it was on the original carrier wave. Now that the set has a single known carrier frequency to deal with, it makes subsequent processing that much simpler. First the signal is passed though a narrow band-pass filter. This cuts out signals at nearby frequencies. Then it is amplified with the advantage that this IF amplifier may be designed to give its best performance with signals at 455 kHz. We are now at the stage of the RF amplifier in Figure 22.7 and, from this point on, the stages of detection, demodulation and AF amplification are as before. One further modification is that part of the detected signal may be used as feedback (p. 171) to increase the gain of the receiver when the received signal is weak. This automatic gain control (AGC) compensates for differences in the strengths of different stations and for the effects of fading.

Television

Like the cinema, television makes use of the persistence of vision. When light rays enter the eye and strike the retina at the back of the eye, the sensation they cause does not cease immediately but remains for some time. The sensation persists long enough for separate, motionless pictures following one another in rapid succession, to give the impression of continuous motion. In television a succession of still pictures are photographed, transmitted and received at the rate of 25 pictures a second. To the observer, the picture on the screen of the TV set appears to be moving in a life-like fashion.

The scene before a TV camera usually contains very many colours, and the same applies to the image that finally appears on the TV receiver screen. Fortunately it is not necessary to transmit details of all these colours. This is because, as far as the human eye is concerned, all colours can be made up by combining lights of three primary colours in differing proportions. The three primary lights are red, green and blue. Red and green combined in roughly equal amounts look to the eye exactly the same as yellow light, even though no light of the yellow wavelengths is present. Red and blue give purple; blue and green give cyan or turquoise. Combinations of all three produce a wide range of subtle colours. Combining all three in equal amounts produces white. All that is necessary for a full-colour TV picture is to transmits the relative amounts of the three primaries in different parts of the picture.

Before transmission, each still picture must be converted to a series of electrical signals. This is done by scanning. To see how this works, let us look at a typical TV camera, the vidicon colour camera (Figure 22.12). First of all see how this camera detects brightness. The lens focuses an image of the scene on to the end of the vidicon tube. This has a transparent conductive layer at the front, with an array of tiny dots of a photoconductive material (such as lead oxide) on its rear surface. An electron gun at the rear of the tube produces a beam of electrons. There are coils to focus the beam and electrically charged plates (as in a CRO, p. 163) to deflect the beam. The beam is made to sweep in horizontal lines across the tube, repeating from top to bottom. As it scans across a line it falls on each of the dots in that row, one after another. Electrically each dot acts as a capacitor in series with a resistor (Figure 22.13). As the beam strikes each dot the capacitor is charged. While the

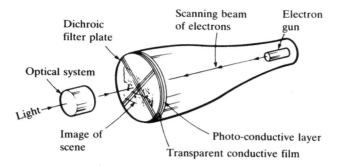

Figure 22.12 *A Vidicon colour TV camera tube, with the beam control system omitted*

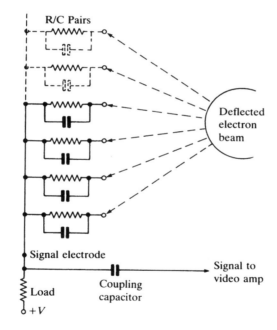

Figure 22.13 *Equivalent circuit of a photoconductive TV camera, showing the resistor-capacitor elements*

beam is scanning other dots, some of the charge leaks away through the resistor (the photoconductive material). The rate at which this happens depends upon how much light is falling on the dot. If the dot is in a bright part of the picture, resistance is low and much of the charge leaks away. If the dot is in a dark part of the picture, resistance is high and only a small amount of charge leaks away.

The next time the beam scans the dot, the amount of current that flows from the beam to the dot depends on how much charge remains there from the previous scan and, therefore, how much needs to be replaced. As each dot is scanned, a signal of amplitude proportional to the replacement current passes to the conductive layer (the signal electrode) and through a coupling capacitor to the video amplifier. The result is a waveform corresponding to the brightness of each dot of each line in the scan. In practice, the lines are not scanned from top to bottom in sequence. If this were done the phosphors at the top of the screen of the TV set would lose much of their brilliance before the scanning had reached the bottom of the screen 1/25 second later. This would result in an unpleasant flickering effect. In practice the tube is scanned twice, the odd lines on the first scan and the even lines on the second scan. Each scan takes 1/50 second, so the complete scan takes 1/25 second. But scanning alternate lines (interlacing) in 1/50 second gives less time for the phosphors to lose brilliance, and this reduces the flickering effect considerably.

Colour information is obtained from the vidicon camera by mounting a filter plate in front of the tube. This bears a pattern of red, green and blue colour filters. The filters may take the form of narrow vertical stripes or some other arrangement. The transparent conductive layer is divided into three sets of striped or other segments to match the pattern of the filters so that individual dots receive either red, green, or blue light and three separate colour signals are obtained.

Figure 22.14 shows the main stages of TV transmission and reception. The information that has to be transmitted includes the following:

- Synchronizing pulses at the beginning of each frame
- Synchronizing pulse at the beginning of each line scan
- During the line scan: luminance signal: the overall brightness as perceived by the eye. This is obtained by mixing the red, green and blue signals in a fixed proportion. This signal is used for brightness control in monochrome receivers.
- During the line scan: chroma signal (chrominance) gives colour information. A complex circuit known as an encoder modulates this and other information on to the carrier wave. Amplitude modulation is used. The sound signal is transmitted separately and is frequency-modulated.

Reception

The receiving aerial is a dipole with a reflecting grid behind it and numerous director elements (Figure 22.15). Returning to Figure 22.14, we can see that the processing of the signals follows the usual sequence for radio reception with the additional feature of detecting synchronizing pulses and using these to keep the scanning in the receiver in step with the scanning in the camera. The tube of a

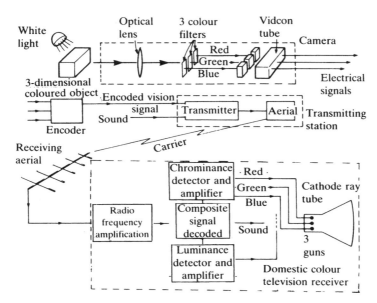

Figure 22.14 *Block diagram of television transmitter and receiver*

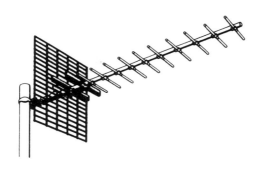

Figure 22.15 *An aerial for reception of TV transmissions in fringe areas*

colour TV is shown in Figure 22.16. It operates in a way similar to the cathode ray tube (p. 163) but has three electron guns, one for each colour. The screen bears dots of three different phosphors, for red, for green and blue. In the tube, immediately behind the screen is a perforated metal aperture plate or shadow-mask. It is arranged so that the three converging electron beams passing through an aperture spread out on the other side of the plate and fall on their respective phosphor dots. From the normal viewing distance the eye

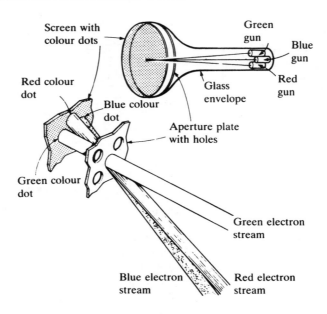

Figure 22.16   *The action of the three-gun colour TV tube*

can not distinguish the individual dots so the light coming from them appears to be merged. Depending on the relative intensities of the three electron beams the screen appears to have a particular colour at that point. For example, if the red and green beams are strong and the blue beam is off, the screen appears to be yellow at that point.

Video recording

The video and sound signals may be recorded on magnetic tape in the same way as in an analogue tape recorder. The main difference is that video signals include frequencies up to 5 MHz so a very high tape speed is required. Given an ordinary audio tape-head, the tape speed would need to be over 60 metres per second. Other methods of recording and reading the tape have been devised, the most successful and popular one being the VHS system known as helical scan (Figure 22.17). The tape is wide and is wrapped round a drum as it passes from reel to reel in its cassette. In a two-head machine, the drum has two record/playback heads spaced 180° apart and the tape is wrapped around half of the drum. The tape is wrapped slantwise so that, as the drum is rotated, one of the heads traces a slanting path on the tape (Figure 22.18). As this head reaches the end of its path the other head reaches the beginning of its path and

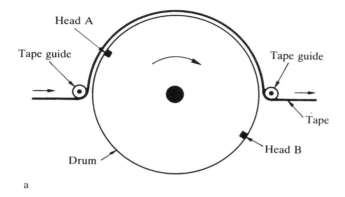

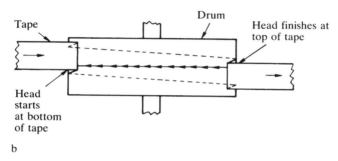

Figure 22.17 *Video-tape wrapping for a two-head drum (a) Seen in plan view (b) Seen in front view. The arrowed line is the path of the heads when they are not in contact with the tape (when in contact they are out of sight behind the drum)*

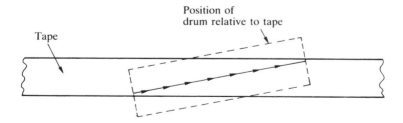

Figure 22.18 *Showing how a diagonal video track is laid on the tape*

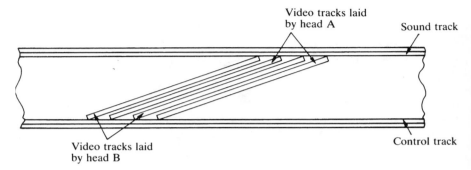

Figure 22.19   *Video-tape track format*

traces the next track. The result is a series of video tracks set diagonally across the tape (Figure 22.19). Additional heads of the deck lay down a control track with timing pulses and a track for the sound. In a stereo recorder there are two narrower sound tracks.

Video cinema

A feature film on video tape costs less than a 35 mm optical print, and the cost of equipping a cinema with a tape-player and projector is less too. A number of cinemas are now equipped with video-tape systems and such systems are widely used for in-flight entertainment. The chief concern is how to provide a picture that is large enough to be viewed by an audience, yet has the brightness and definition comparable to that produced by an optical projector. The **light guide projector** employs three tubes of the type shown in Figure 22.20. Each produces

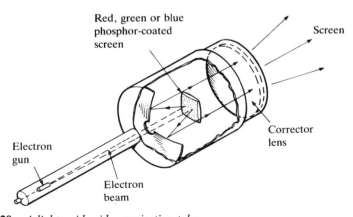

Figure 22.20   *A light guide video projection tube*

its own coloured image (red, green, or blue) and these images are superimposed on the screen to form a single full-coloured image. Since the light from three tubes is combined, and there is no reduction in brilliance due to shadow-masks, a bright image is obtained.

The optical cinema owes its brilliant image to the exceedingly bright lamp that is used as the source of light. The phosphor image on a video tube can not match this level of brilliance. One solution to this problem is found in the **Eidophor** tube (Figure 22.21). Its source of light is a high-powered xenon lamp. The light is focused on to a mirror that is covered with a film of special oil. When the electron beam is not scanning, the surface of the film of oil is smooth. The light reflected from the mirror is blocked by the bars and does not reach the screen. When the film of oil is scanned, its surface is disturbed. The higher the intensity of the beam, the greater the disturbance, and the more light is reflected to pass between the bars to the screen. For colour projection, three Eidophor tubes are used, one for each primary colour. The projected images are superimposed on the screen to obtain a full-colour picture.

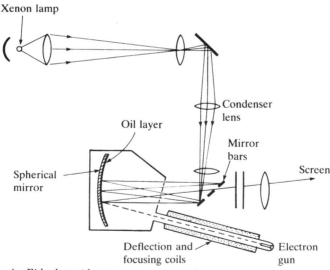

Figure 22.21   *An Eidophor video projection tube*

23

Microwaves

Microwaves are electromagnetic waves (p. 237) with frequencies ranging from about 3 GHz up to about 300 GHz. The higher the frequency, the shorter the wavelength so that, at the top end of the microwave range, the wavelength is only a few millimetres. This has important effects on the way microwaves behave, as will be seen later. The range of wavelengths of microwaves puts them between the electromagnetic radiation used for broadcasting radio and TV programs and the waves of infra-red. This is why microwaves are similar to radio waves in some respects but also show some of the features of infra-red.

Generation of microwaves

As explained on p. 236, electromagnetic waves are generated by the rapid to-and-fro movement of charged particles, typically of electrons. The frequency of the radiation depends on the rate at which the electrons oscillate. Generating microwaves is just a matter of building an oscillator with the required high frequency. Oscillators of the kinds used in ordinary radio transmitters are not able to oscillate at microwave frequencies with sufficiently high power output. This is mainly because the dimensions of many of the components are greater than the wavelengths. For example, tiny capacitances that have no effect at broadcast radio frequencies exert an unduly high influence at microwave frequencies. The whole concept of oscillator and transmitter design has to be thought out again, particularly, the kinds of components used. Some of the special techniques used for microwaves are described on p. 256, but first we look at a well-established microwave generator, the **magnetron** (Figure 23.1). The main part of this is the anode, which consists of a copper block which has a large central cylindrical cavity surrounded by (usually) eight smaller cylindrical cavities. Running down the central cavity is the cathode, made of metal and heated by an internal coil. The whole magnetron is enclosed and its interior is a vacuum. A strong permanent magnet outside the cavity produces a magnetic field directed along the axis of the magnetron. In operation the cathode is made negative and the anode made positive so that there is a radial electrical field (from anode to cathode) in the cavity of the magnetron. We thus

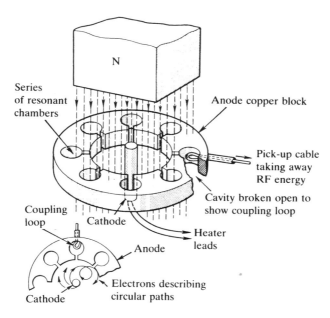

Figure 23.1 *A magnetron microwave oscillator*

have the situation known as a **cross-field** because, at all points, the magnetic field is at right angles to the electrical field.

The hot cathode emits electrons from its surface, and these are repelled by the negative charge on the cathode. The electrons are accelerated radially toward the anode by the electric field but, at the same time, they are accelerated by the magnetic field in a direction at right angles to the magnetic field. The relative strengths of the fields are such that they describe circular paths, gradually moving toward the anode.

The eight outer cavities function as resonating circuits. Figure 23.2 shows a resonating circuit constructed from an inductor coil and a capacitor (p. 62). This resonates at a fixed frequency, depending on the value of L and C. If the inductance is reduced to that of a single turn of a coil (the wall of the cavity), and the capacitance to a narrow gap between two plates as shown in the figure, the resonant frequency is very high, in the microwave range. By machining the cavities to the correct dimensions we produce a cavity that is tuned to oscillate at any required microwave frequency. Initially, owing to the random distribution of electrons around the cavities, a weak oscillation at this frequency occurs, causing an electromagnetic field across the aperture of the cavity. The fields of adjacent cavities are in opposite directions at any instant (they are 180° out of phase), but all fields are reversing in direction at the microwave

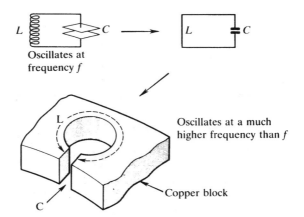

Figure 23.2 *Showing how the cavities in a magnetron are each equivalent to an inductor-capacitor resonant circuit*

frequency. As electrons from the cathode come nearer to the anode they are affected by these electromagnetic fields, being alternately accelerated and decelerated by the weak fields. An accelerating electron gains its energy from the field between cathode and anode, but a decelerating electron transfers its energy to the electromagnetic field of the cavity. Eventually a cloud of electrons moves around the cavity close to the anode; because of the alternate acceleration and deceleration, the cloud becomes bunched. Its decelerating regions come opposite the apertures just at the right moment to transfer energy to the electromagnetic field and strengthen it. As a result, oscillations at the microwave frequency gradually increase in amplitude and the fields become stronger. In this way, the energy from the cathode-anode field is converted to electromagnetic energy at microwave frequency. Electrons gradually lose energy in this process and, after circulating the cavity a few times eventually become absorbed by the anode.

The electromagnetic energy, which from now on we may simply call **microwaves**, is taken from one of the cavities by means of a wire loop inserted in the cavity. Next, the microwaves may be amplified by a **cross-field amplifier** which is very similar in construction to a magnetron. The main difference is that two metal strips (known as straps) run round the anode, connecting the walls between the cavities in alternate pairs (Figure 23.3). These ensure that the oscillations in adjacent cavities stay exactly 180° out of phase. Microwaves from a magnetron are introduced into one of the cavities by means of a wire or loop connected to a coaxial cable (like the cable used to connect a TV set to its aerial). Action very similar to that of the magnetron causes the amplitude of the oscillation to increase and the amplified microwaves are extracted from one of the other cavities by another coaxial wire or loop.

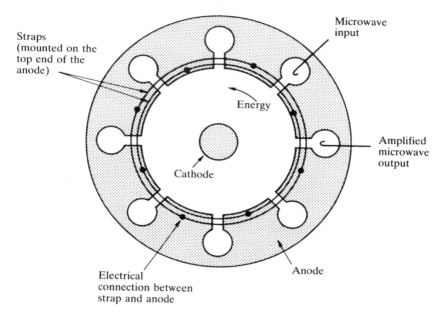

Figure 23.3 *A cross-field microwave amplifier*

Waveguides

Microwaves may be conducted by coaxial cables, which are convenient for many purposes, but they are more often conducted by rather more unusual means, such as waveguides (Figure 23.4). A waveguide is a metal tube, usually rectangular in section. We can think of the waves as passing along inside the tube, being reflected off the walls of the tube, zig-zagging from one end of the tube to the other. This is similar to the way in which light is conveyed along inside a glass rod or optical fibre by total internal reflection (p. 236). The advantage is that the microwave is confined to the interior of the waveguide. It does not radiate from it and cause interference with surrounding equipment, and there is very little loss of energy, so transmission is very efficient, especially at the highest frequencies. Note that the waves are passed along the waveguide in the air-filled cavity, not conducted in the metal walls, which are there solely to reflect the waves. Depending on the dimensions of the wave-guide and the wavelength of the microwaves, there are various modes of transmission. One mode commonly used is the TE10 mode shown in Figure 23.4, in which the guide is exactly half a wavelength wide.

Microwaves are introduced into a waveguide by inserting a coaxial cable in a cavity of the magnetron or amplifier and running it into one end of the guide.

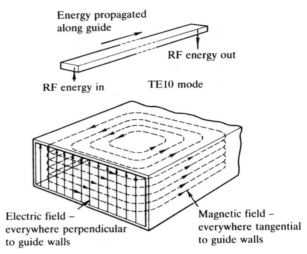

Figure 23.4   *The magnetic and electric fields inside a waveguide when it is conducting microwaves*

A similar cable at the other end of the guide may be used for collecting the microwaves.

Microstrip

Microstrip is a more recently developed technique for microwave circuit construction. The circuit is formed by metal (often copper or gold) conducting strip on a non-conducting substrate such as alumina. The conductors are formed on one side of the board, the other side of the board having a continuous film of metal to produce a ground plane. Figure 23.5 shows the electrical field when a microwave signal is passing along a microstrip transmission line. The strips may be formed by etching the patterns on one metallized surface of the board, in a manner similar to that used for producing PCBs (p. 134). Another technique is to screen-print the circuit on uncoated board, using thick conductive ink. Microstrip circuits can also be fabricated as integrated circuits using gallium arsenide (p. 143).

The fact that the dimensions of the strips (a few millimetres wide) are of the same order as the wavelength of the microwaves gives rise to effects which have useful applications. For example, if the width of the conductor changes suddenly at a given point, the electromagnetic field is distorted (Figure 23.6). The effect is the same as if there was an inductor-capacitor network at that point. In other words, simply by varying the width of the strip we are able to 'construct' passive components on the circuit board. In Figure 23.6 we have the equivalent

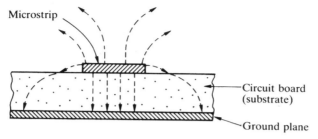

Figure 23.5 *A vertical section of a microwave circuit board with a microstrip transmission line*

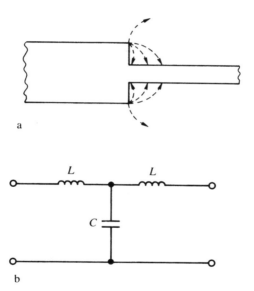

Figure 23.6 *Showing how microstrip can have the properties of components (a) Plan view of a length of microstrip showing distortion of the electromagnetic field where the strip is suddenly narrowed (b) The equivalent circuit*

of a low-pass filter. Other effects are obtained by using parallel strips with a gap between them, the coupling of fields between the strips having the effect of an inductor and capacitor in series. The circuits are 'tuned' to any required frequency by their widths, the lengths of each section and the spacing between adjacent sections. This technique can be applied to create the equivalents of other components such as transformers and mixers. Figure 23.7 shows a capacitor formed by interdigitating ends of two microstrips. Although the capacity of such a device is small, it is large enough to be effective at the very high frequencies of microwaves.

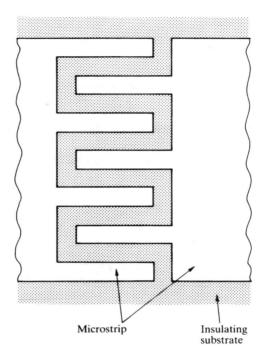

Microstrip   Insulating
                    substrate

Figure 23.7   *Obtaining the equivalent of a capacitor with microstrip*

Microstrip is used in conjunction with other special devices suited to high frequency operation. **The varactor diode** is a microwave version of the varicap diode (p. 73), and is often used as a tuning capacitor in transmitters and receivers. Another technique for tuning depends on using a sphere of a ceramic material. Essentially the material consists of yttrium and iron oxides formed into a garnet material and are therefore known as **YIGs**. The sphere, which is between 0.2 mm and 2 mm in diameter is placed close to a narrow loop of microstrip, and is exposed to a steady magnetic field, produced by an electromagnet. When microwaves pass along the strip they induce electromagnetic vibrations in the YIG at the same frequency. At a fixed frequency (depending on the strength of the applied magnetic field) a resonating field is set up in the YIG. This induces the microwave circuit to oscillate at the same frequency. The resonant frequency is set by the strength of the field produced by the electromagnet. By adjusting its field strength electrically we are able to tune the microstrip circuit to resonate at the required frequency. YIGs produce precisely-tuned frequencies.

Discs or cylinders of ceramic materials such as barium titanate and zirconium titanate are coupled to microstrip circuits to be used as resonators in a similar way to YIGs, except that they do not depend on a magnetic field for tuning. The resonant frequency depends upon the dimensions of the disc and to a certain extent its distance from the microstrip and from the walls of the enclosure. A movable plate mounted on the wall of the enclosure may be used to tune the resonator over a limited range of frequencies.

Microwave communications

A microwave communications transmitter consists of a microwave generator, producing a carrier wave which is then modulated by the audio or other signal (for example, computer data). The principles of modulation and demodulation at the receiver are similar to those employed in ordinary radio (p. 238). Frequency modulation is preferred.

Microwaves are transmitted into the atmosphere by using a dipole aerial (p. 240) of suitable dimensions. This type of aerial may also be used at the receiver. Dipoles, in which the antenna is approximately half a wavelength long, are suitable for microwaves of lower frequencies, such as a few hundred megahertz. Part of the radiation travels over the Earth's surface (**ground wave**) and part is projected upward (**sky wave**). Although a dipole works well with transmission at ordinary (long wave, medium wave and short wave) frequencies, the ground wave rapidly becomes attenuated at frequencies greater than about 1 MHz, so reception by ground waves is restricted to distances of a few tens of kilometres. Dipoles are used for the transmission of broadcast radio and TV signals, including transmissions to mobile telephones. Because of the limitations of distance, numerous TV and radio stations are set up throughout the country to cover local areas. Similarly, the country is divided into cells, each with its own transmitter to communicate with mobile telephones. The sky waves from a dipole are of little use for communications because, unlike normal radio wavelengths, microwaves pass through the ionosphere (p. 239) and are lost into space.

For higher frequencies we use a parabolic reflector. A waveguide conveys the microwaves to the focal point of the reflector. They emerge and are focused into a more-or-less parallel-sided beam. At the receiver, a similar parabolic reflector concentrates the arriving waves into another waveguide which conveys the waves to the receiving circuit. The advantages of the reflector antenna is that it allows transmission over relatively long distance with minimum loss of power. Transmission is by line of sight so transmitter and receiver must normally be situated on high ground, with the antennae mounted on towers. Since the beam is highly directional, only those receivers in line with the beam can receive the signals. But high-frequency waves are attenuated by water droplets in the atmosphere, such as fog and rain, which further limits transmission distances. To cover long distances, it is usual to set up a chain of repeaters which receive a signal, amplify it, and then re-transmit it to the next station along the line at a slightly different frequency. Nowadays, with the improved performance of optical fibre transmission (which does not require line-of sight, and is not subject to outside interference or climatic attenuation) microwave links are being replaced by optical fibre links (p. 235).

Microwaves are of particular importance for satellite communications since transmissions at frequencies of 3 GHz or more are not reflected by the ionosphere. TV and other transmissions are directed up to a 'stationary' satellite, using a parabolic reflector to produce a narrow beam of radiation. The satellite is not actually stationary. It is in orbit above the equator, but its orbiting time is exactly 24 hours, so keeping the satellite in position above a fixed point on the Earth's surface. We say that it is in **geostationary orbit**. From its high altitude, several tens of kilometres above the Earth's surface, the satellite can be 'seen' from a little less than half the Earth's surface. It can also 'see' the same area. The transmission is directed up to the satellite from the transmitting Earth Station (Figure 23.8) and received by the satellite. It is amplified and then re-transmitted on a slightly different wavelength, to avoid interference between the incoming and outgoing signals. For telecommunications, a narrow downward beam may be used but, for TV broadcasts intended for direct reception by viewers over a country-wide area, the downward beam is wider. On the ground, the signal is received by a parabolic reflector and fed to the receiver circuit. In addition to the main telecommunications channels, the satellite is in microwave contact with its ground station, which receives data concerning the status of the systems inside the satellite and its location in space. The ground station can also transmit instructions to the satellite, including

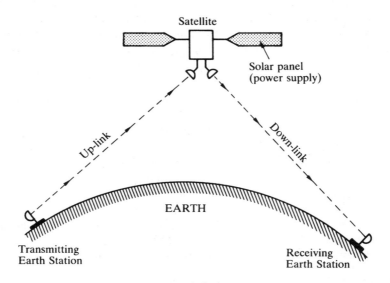

Figure 23.8   *Satellite telecommunications link*

instructions to power up its small rocket motors to correct its attitude from time to time.

With three geostationary satellites, it is possible to cover the whole of the Earth's surface except for the polar regions. Unlike cable links, the cost of satellite links is independent of the distance. In underdeveloped areas, where reliable land lines are scarce (or non-existent) and costly to install, satellite communications bring great advantages.

Radar

Radar (or **ra**dio **d**etection **an**d **r**anging) was first demonstrated in 1935 by Sir Robert Watson-Watt. Subsequently it became of major importance for detecting aircraft in wartime, a fact which did much to accelerate its development. The principle is simple. A transmitter sends out a series of very short pulses of microwaves. These travel outward from the antenna, being reflected off any object that they strike. Part of the radiation is reflected back to the location of the transmitter, where there is a receiver. Electronic circuitry measures the time passing between transmission of a pulse and the reception of its echo. The time taken is proportional to the distance of the object, so the distance of the object can then be calculated.

Although radar can be used solely for measuring distances, it is more often used for measuring direction as well. The antenna is made to rotate continuously on a vertical axis so that it scans the surrounding area repeatedly. One form of antenna, often used on ships for detecting other surface craft at night and in fogs, is shown in Figure 23.9. The waveguide conveys pulses of high-energy microwaves from the amplifier directly to the antenna horn. With the typical cheese-dish antenna the vertical height of the beam is large, but this is

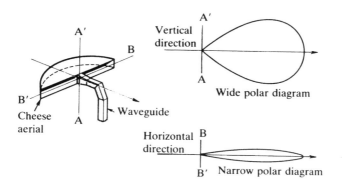

Figure 23.9   *A cheese aerial and its polar diagrams*

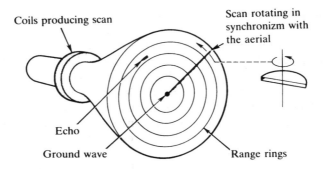

Figure 23.10   *A typical PPI display*

not of significance. The beam is narrow in the horizontal plane, making it easy to pin-point the direction of the detected object. In aircraft radar systems such as are in use at airports a huge rotating array of parabolic reflectors directs a radar beam up into the sky. Most often the information obtained from the rotating antenna is displayed on a *plan position indicator* (Figure 23.10). This consists of a cathode ray tube with electrodes arranged so as to produce a narrow fluorescing line which rotates about the centre of the screen. Its rotation is synchronized with that of the antenna so that if, for example, the antenna is directed north, the line is directed vertically up the screen. As the line is scanned from the centre to the margin of the screen, it displays reflected signals as an increase in brightness of the line. The distance of the bright spots from the centre of the screen is proportional to the distance of the reflecting objects from the radar station. There is also a bright spot at the centre of the screen due to the reflection of the ground waves (see above) from nearby buildings and other structures.

Instead of the PPI the display may be a computer monitor. This allows for more sophisticated processing of the information and for the simultaneous display of ancillary information on the screen. For example, the computer can distinguish between stationary objects that show up on every rotation of the antenna and those which have moved since the previous scan. This may be useful in certain situations, for example to pick out the movements of vessels in a harbour.

Sub-surface radar

Microwaves are able to penetrate beneath the surface of the soil and reveal the presence of buried objects. Under favourable soil conditions, objects as far down as 50 m may be detected. The equipment consists of a transmitter to send pulses of wide-band microwaves into the soil and a receiver to detect radiation reflected back from buried objects. Whereas in the case of conventional radar the transmitter and receiver are stationary and the detected object is moving, with sub-surface radar the transmitter/receiver is moved in a regular pattern across the area being scanned and the objects themselves are still. The signals from the receiver are fed to a computer for analysis. This produces a display showing a vertical section through the strip of soil currently being investigated. This may be in colours that distinguish the different types of material present below the soil. If several scans are carried out along parallel grid lines, a picture of the sub-surface features of an area of terrain may be built up.

Applications of sub-surface radar include finding buried cables and pipes, locating persons buried under snow avalanches, and in archaeological surveys. One researcher using SSR has discovered a boat buried high in the mountains of Turkey; this is possibly the remains of Noah's Ark.

Doppler systems

These are used for detecting moving objects. Perhaps one of their most familiar applications is for automatically opening the doors of a shop whenever a customer approaches. The Doppler Effect is noticed in everyday life if we are standing on the pavement when a fire-engine rushes by at high speed. As the vehicle passes, the apparent pitch of its siren abruptly falls. The frequency of the sound is higher when the fire engine is approaching, and lower when it is receding. A similar effect is found with reflected microwaves. Microwaves of a given frequency are sent from a transmitter mounted above a door. If a person is approaching the door, some of the waves are reflected back toward the transmitter and have a slightly higher frequency than the transmitted waves. The change of frequency is only very slight, but it is measurable. For example a person walking toward a transmitter at 6 km/h causes the frequency of a 10 GHz microwave beam to increase to 10.000 000 056 GHz. This is an increase of only 56 Hz. To detect this increase the detector circuit mixes the received signal with part of the transmitted signal. The two signals interfere with each other to produce a beat frequency of 56 Hz (p. 242). This is detected electronically and triggers a circuit which opens the door. Note that this action depends on the person being in motion. If two persons are standing talking outside the door, but are not moving, the reflected signal has the same frequency as the transmitted signal and there is no beating.

A similar principle is used on some car alarm systems. The transmitter/receiver is usually mounted on or near the floor of the vehicle, the microwaves being reflected by the metal floor to give a wide area of coverage. Any moving object inside the car causes a shift of frequency, which is detected and triggers the alarm to sound. It is also possible to measure the rate of beating and thus find the velocity of the moving object. This is the basis of the devices used by police to measure the speed of vehicles, using equipment positioned at the roadside.

Microwave ovens

Microwave energy is readily absorbed by the atoms and molecules of any non-conducting material it passes through and is converted into heat (thermal) energy. The microwaves behave in much the same way as their nearest relatives in the electromagnetic spectrum, infra-red radiation. In microwave ovens the frequency (usually 2.45 GHz) is such that it is efficiently absorbed by water molecules, which are nearly always present in high proportions in the food. The microwaves are generated by a magnetron and conducted by a wave-guide into the cooking section of the oven (Figure 23.11). The cabinet of the oven is made of metal, so it acts as a totally enclosed wave-guide, confining the microwaves to the oven. A switching mechanism ensures that it is not possible for the magnetron to be activated while the door is open. Although the door has a transparent panel to allow inspection of the food while cooking, this panel is covered with a screen of perforated metal to reflect microwaves back into the interior.

At the point where the microwaves enter the cooking section there may be a slowly turning fan or stirrer to continually alter the otherwise stationary micro-wave fields in the oven to promote even cooking. For the same reason the food is placed on a rotating turntable in all but the cheapest ovens. Microwaves penetrate food and cooking containers easily so that heating occurs more-or-less evenly throughout the food. This appreciably reduces cooking time when

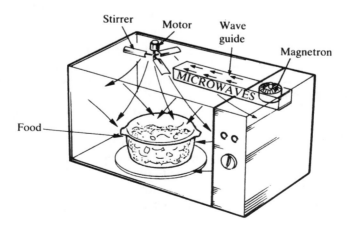

Figure 23.11   *A microwave oven*

compared with a traditional thermal oven. It makes microwave ovens (usually with a 600 W to 700 W magnetron) much more convenient and economical to operate than a conventional electric oven, whose heating elements may require 3 kW or more for much longer periods. In addition, the walls of a microwave oven are made of metal, so they do not absorb microwave energy. After a period of cooking the walls are only slightly warmed by convection from the cooking food. By contrast, the walls of a conventional oven become very hot indeed, a source of accidental injury. One slight disadvantage is that cooking with microwaves does not readily produce the crusty or roasted outer layer that is familiar on cakes, roast joints and similar foods cooked in a conventional oven, but there are ways of producing this effect by other means.

24
Medical electronics

Electronics is playing an ever-increasing role in medicine. Hardly any patient entering hospital for examination or treatment is exempt from the attention of electronic apparatus. Whether it be something simple such as a pocket-sized device for processing a blood sample automatically or a machine for automatically measuring and recording blood pressure and pulse rate at regular intervals, it seems as if every piece of equipment has a microprocessor or logic circuit hidden away inside it. Even in the home, inexpensive electronic devices for measuring pulse rate and blood pressure are becoming commonplace.

Body scanning

One of the most dramatic advances in medical electronics has been in the field of body scanning. The EMI **whole body scanner** is illustrated in Figure 24.1 The scanner consists of a source of a narrow beam of X-rays and an X-ray detector that are located opposite each other and able to move round a circular path. The body of the patient lies at the centre of the circle. The detector measures what fraction of the X-rays has been absorbed while passing through the body. This information is fed to a computer and stored in its memory. As the appa-

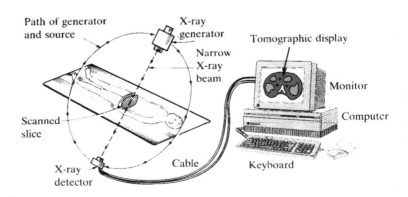

Figure 24.1 *Computer-assisted tomography; the principle of the EMI whole body scanner*

ratus rotates the 'slice' of the body that is being scanned is penetrated by X-rays from all directions. Parts that are in the shadow of bone when the beam comes from one direction are clear of shadows when the beam comes from another direction. The computer is programmed to process the stored information and produce a display that shows the shapes and positions of the structures present in the slice. Early versions of the EMI scanners were small and used only for scanning the brain. They can locate brain cysts and tumours in scans lasting only a few minutes. More recent equipment can scan the whole body.

Another technique for body scanning is **magnetic resonance imaging** (MRI). Scans of slices though parts of the body, especially the brain, are made by detecting the resonance of atoms in a strong magnetic field. This allows maps to be made according to one of a number of tissue characteristics, such as proton density. The technique requires a very strong and uniform magnetic field throughout the part being scanned. Electromagnets employing superconductors (p. 27) are excellent for producing the fields required. Protons in water and tissues align themselves in the field. A burst of radio-frequency radiation is used to excite the protons, which then release energy as they return to their unexcited state. This energy is detected and the measurements fed to a computer which uses the information to build images of body structure. MRI has an advantage over other scanning techniques that it does not expose the subject to ionizing radiation, such as X-rays or emissions from radioactive substances.

Another application of MRI is **projection angiography**. This relies on the fact that the faster the flow of blood the brighter the image on the MRI screen. This technique is used to measure rates of blood flow in regions of the body such as the head and neck. Information gained in this way is useful in the treatment of strokes and cerebrovascular diseases.

LASS

This is the name for the Loughborough Anthropomorphic Shadow Scanner, which is under development at the time of writing and promises many useful applications. The human subject, clad in a body stocking, stands on a turntable. Four banks of four projectors produce four narrow bands of light running vertically down the subject. The subject is scanned by two banks of seven TV cameras placed on opposite sides of the subject. The turntable is slowly rotated and the positions of the bands registered by the cameras. The data from the cameras is analysed by a computer to build up a three-dimensional model of the subject in terms of the radial coordinates of points on the body surface.

In the medical field, this equipment has applications in connection with X-ray treatment for diseases such as breast cancer. It is essential that X-rays should not be directed towards sensitive parts of the body, such as the bone marrow. LASS gives an accurate model of the subject's body in three dimensions. This is of great use in positioning the X-ray tubes so as to localize the radiation where it is required and avoid regions which must not be exposed. The equipment can also be used to collect data on samples of the general population to establish typical body proportions. Such information is of use to health scientists and also clothing manufacturers, furniture designers and bio-engineers.

Measuring bodily potentials

Electronic equipment is used to measure the small electrical potentials developed in the body as a result of the activity of muscle and brain tissues. By monitoring these potentials it is possible to detect abnormalities in bodily functions and so diagnose diseases. The heart is a powerful muscle and as it beats it generates potentials which spread throughout the body. The physician can measure and record these potentials by means of electrodes connected to the patient's limbs. The potentials conform to a pattern that is the same in all normal people so that divergence from the pattern indicates some type of abnormality. This technique is known as **electrocardiography** (or ECG for short). Figure 24.2 shows ECG measurements being made on a patient and a typical recording. The potentials generated by the heart are picked up by silver or platinum electrodes attached to the wrists and left leg of the patient. Measurements are made in turn of the variations in potential difference between the right arm and left arm, the left arm and left leg, and the right

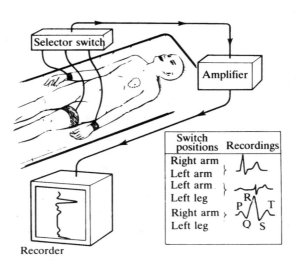

Figure 24.2   *Electrocardiograph equipment*

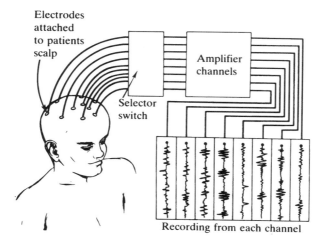

Figure 24.3   *Electroencephalograph equipment*

arm and left leg. Because the body is a high-impedance source (that is, the potentials are measurable but are unable to deliver a large current to the apparatus) the amplifier has FET inputs, which have high impedance (p. 76).

The output of the amplifier is applied to a pen recorder which makes a permanent record of the signals from the patient. This is known as an electro-cardiograph. The usual pen recorder is not able to follow signals exceeding 100 Hz in frequency so special recorders able to respond to 1 kHz signals are used for ECG recordings.

Potentials in the brain are explored with similar equipment to produce an electroencephalograph (Figure 24.3). High-gain amplifiers are needed as the brain is less physically active than the heart and produces potentials of only about 100 microvolts. The potentials are detected in the patient's scalp by using small disc electrodes firmly attached to the skin by a harness. It is usual to have up to 24 electrodes. Each provides a series of signals which are fed to separate amplifiers and pen recorder units. The signals can also be displayed simultaneously on a computer monitor.

Stimulators

The use of electronics to stimulate reactions in patients has most applications in the field of cardiac medicine. The heart beats as a result of a nervous impulse which is conducted through the entire heart, beginning in the auricles and ending in the ventricles. This impulse is electrical and can be imitated by electronic means. It should not be confused with the potentials measured by an ECG, which are the *result* of the heart beating. In some heart diseases a temporary or permanent disorder of the conducting mechanism occurs so that the heart beats too slowly or irregularly. Stimulators may be applied either to the chest wall or internally to the heart muscles. The stimulator produces a spike-shaped waveform at a frequency suitable for making the heart beat with a regular rhythm. In chronic disease of this type a small semiconductor stimulator or **pacemaker** is inserted in the body by a surgical operation. It lies among the muscles in the region of the shoulder. Leads from the pacemaker pass to electrodes inserted in the heart muscles. Electrical spikes are delivered to the heart muscles so that they contract regularly and at the correct rate. The energy to power the pacemaker comes from a small battery or from a tiny nuclear-powered device in the pacemaker. This relies on the energy emanating from radioactive plutonium-288. It can remain operating in the body for ten years or more without further attention.

Another serious heart condition is heart flutter or fibrillation. In this condition, the heart works normally for most of the time but occasionally gets out of control and flutters wildly. Such a condition is fatal unless the normal beating of the heart is rapidly restored. The method of doing this is to administer a relatively large electric shock to the heart. This stops the heart for an instant, after which it generally resumes its normal action. When a person is affected, the shock may be given from a pulse generator held against the chest. Alternatively, a miniature defibrillator is permanently implanted in the body. It is connected to electrodes implanted in the heart muscles. It monitors the normal heart beat and, if it detects fluttering, it automatically delivers a shock to the heart within a few seconds. Prompt correction of the heart's irregular action by this device has saved many lives.

Microwave diathermy

It is explained on p. 264 that microwaves cause heating when they pass into materials containing water molecules and other molecules that are capable of absorbing their energy. Such materials include the tissues of the human body, into which microwaves are able to penetrate several centimetres. Under carefully controlled conditions, microwaves are used to heat parts of the body, particularly deep-seated muscles, as part of physiotherapy. To prevent damage to the eyes, both patient and operator wear wire-mesh goggles.

Electronics helps the disabled

There are numerous examples of the ways in which computers (which owe their existence to electronics) have been brought to the aid of the disabled. Text-to-speech programs enable the blind to 'read'. Speech-to-text programs enable the deaf to 'hear'. Special computer input devices consisting perhaps of a few microswitches operated by suitable parts of the patient's body, allow the physically incapacitated to initiate a range actions that would be otherwise quite impossible.

On a less highly technical level, but with benefit to a greater number of people, the installation of induction loops in theatres and cinemas is of great help to the partially deaf. An **induction loop** consists of a coil of one or a few turns of wire running around the walls of a room. The loop is connected to an audio amplifier in place of the loudspeaker. In the home, the loop may be connected to the TV set or CD player. The coil generates an alternating electromagnetic field within the room. A hearing aid contains a minute magnetic core surrounded by a pick-up coil. The induction loop, core and pick-up coil act as a transformer. When the hearing aid is switched to the T (telephone) setting, currents generated in the pick-up coil are fed to the amplifier of the hearing aid. This function was originally intended for coupling the hearing aid to the telephone circuit, enabling the wearer to hear telephone conversations more easily, and is still regularly used for this purpose. The T setting also allows the wearer to make use of induction loops. If an induction loop is in operation, signals produced by the loop are heard by the wearer of the aid. Apart from obvious uses such as enabling the wearer to listen to radio and TV sound without the need to wear a headset, the induction loop makes it easier to hear conversation. In any room a listener hears conversation by pressure waves coming directly from the speaker and indirectly by reverberation. At distances from the speaker greater than about 2 m, the reverberation predominates, making it difficult to hear the conversation. If there are several persons speaking in the room at the same time, the difficulty is considerably increased. By speaking into a microphone connected through an amplifier to an induction loop the speaker can, in effect, be brought closer to the deaf person and be more clearly heard. For this reason an induction loop can be of benefit also in a small auditorium and committee room.

Radioactive tracer elements

It is sometimes useful to be able to follow the path of a chemical substance as it passes from one organ to another in the body. As an example, take a patient suffering from a disease of the thyroid gland, which is situated in the neck. We may want to be able to find out how well the gland is making thyroxine, which is a hormone concerned with the proper development of the body. Thyroxine contains the element iodine, which is not present in many other substances in the body. We can give the patient food that contains iodine and see what the diseased thyroid gland does with it. Chemical analysis of tissues or body fluids does not give a complete picture of the process. One of the reasons for this is that chemical tests do not allow us to distinguish between the iodine recently supplied in the food and iodine that has been present in the body for a long time previously.

If the iodine supplied in the food is mixed with a small quantity of radio-active iodine, we can use the radioactivity to distinguish the iodine supplied in the food from iodine already present in the body, or from normal non-radio-active iodine supplied in later batches of food. The radioactive iodine acts as a **tracer**. Analysis is made easier because the radiation from radio-iodine passes through the tissues and we can detect it from outside the body. There is no need to remove tissues or fluids from the body for testing.

The radiation is measured by means of a solid state nuclear radiation sensor (see Figure 11.11, p. 115). When the hospital physician uses this equipment the patient is fed with food containing normal amounts of iodine to which minute and harmless amounts of radio-iodine have been added. The sensor is placed against the patient's throat. After a while the instrument shows that increasing amounts of radiation are coming from this region as the thyroid gland takes up and accumulates iodine from the blood stream. By measuring the changes in the amount of radiation, the physician is able to investigate the behaviour of the diseased gland.

Radioactive tracers are widely used in medicine, in scientific research and in industry. By their use we can detect and follow the movement of exceedingly small quantities of materials, often only a few micrograms. For example, tracer techniques can detect minute leakages in drainage systems, or the wear of metal in the bearings of a car engine.

Radio pill

Radio pills are, as the name suggests, devices to be swallowed. The idea is that they sense the conditions in different parts of the alimentary canal as they pass through them. The information is transmitted by the pill as a radio signal which can be picked up and analysed by the physician while the pill is passing through the body. Before the advent of the transistor it was impossible to make a device small enough to be swallowed without considerable discomfort to the patient. Using a transistor powered by a miniature cell, a pill need be only about 20 mm long and 8 mm in diameter. Figure 24.4 shows a typical radio pill which will sense and transmit information about the pressure, temperature, acidity and several other factors of interest to the physician. The pill is essentially a minute radio transmitter working at about 300–400 kHz. The information is radiated through the body to a receiver outside, which detects the signals and passes them to a recorder. Information is conveyed by varying the frequency of the carrier wave and this can be done in various ways. If, for example, we want to know the pressure inside the intestines, the inductor in the circuit is made to alter the tuning of the transmitter, its iron core being displaced in accordance with the pressure on the walls of the pill. If the acidity

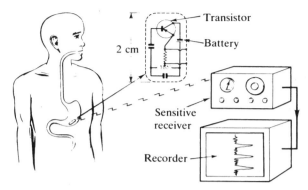

Figure 24.4   *A 'radio pill'*

inside the stomach is to be measured, this can be done by placing pH electrodes (p. 114) on the surface of the pill. When immersed in the acid solution in the stomach, the electrodes generate a p.d. proportional to the acidity. This is used to alter the frequency of the transmitter by altering the potential of the base of the transistor. If temperature is to be measured, the effect of temperature on the characteristics of the transistor can be used to vary the frequency.

Ultrasonics

Ultrasound is sound of such high frequency that it can not be heard by the ear. Frequencies of 16 kHz or higher fall into this category, though frequencies as high as 40 kHz are often used. Ultrasonic techniques have a similarity to radar. The ultrasound is generated by an oscillator running at the required frequency. A piezo-electric (p. 107) transducer, cut so as to resonate strongly at the required frequency, is used to convert the electrical oscillations into ultrasound waves. A similar piezo-electric device is used to detect ultrasound; this is connected to a circuit that indicates the amount of ultrasonic energy being received.

Ultrasonic techniques have a similarity to radar (p. 261). In ultrasonic scanning equipment bursts of ultrasonic energy are applied in a narrow beam to parts of the human body, for example the skull and the abdomen. The ultrasound penetrates the tissues but part of the beam is reflected at the boundary layer between organs, where sudden changes of tissue density occur. The reflected ultrasound is detected and information is fed to a computer. This produces a picture on the screen of its monitor, showing the internal organs of the body as they would be seen in section. The technique provides information which can be used to support or replace examination by X-rays. The ultrasonic scanner is of particular use in examinations during pregnancy.

The frequent use of X-rays in such circumstances could cause genetic damage to the unborn child and to the mother herself, whereas ultrasound is relatively safe.

A further use of ultrasonics in medicine is the ultrasonic hypodermic needle. The needle incorporates a suitable transducer to which is applied the output of an ultrasonic generator. The ultrasonic transducer makes the needle act rather like a road drill. Its advantage is that it can be used for penetrating deeply into body tissues without damaging the surrounding tissues or causing pain.

The sterilization of surgical instruments is another sphere where ultrasonics are playing an important part. The ultrasonic transducer is placed in a sterilization bath together with agents such as detergents to lower the surface tension of the water. When the oscillator is switched on, the micro-agitation it produces in the water removes dirt from the tiniest crevices in the instruments and makes subsequent sterilization more effective. Ultrasonic cleaning baths are widely used for other purposes, for example for cleaning small objects such as the gear wheels of watches and other delicate mechanisms.

25

Electronics in industry

Electronics has shown an increasing number of applications in industry for several decades. Electronic techniques are widely used for measuring all kinds of attributes from the thickness of fabrics to the viscosity of paint. Backed by computers, and often using robots, they can control complex industrial processes from glass-making to paint-spraying of automobiles. Computers can take the place of supervisory staff and skilled manipulators and may also be used to work out costs and profits. This chapter gives a few examples of the industrial use of electronics to illustrate the vital role that it plays in modern industry.

Electronic counting

There are a number of ways in which electronic techniques are used in factories where large quantities of small items, for example small machine parts, have to be counted.

One method uses an ordinary microphone as shown in Figure 25.1. The small metal parts drop singly on to a platform under which a microphone is situated. The microphone detects the sound as each item hits the platform, prior to slipping away down the chute, to be carried away for packing. The microphone output is connected to an amplifier, the output of which is fed to a

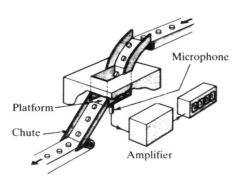

Figure 25.1 *Counting small metal articles by using a microphone and counter*

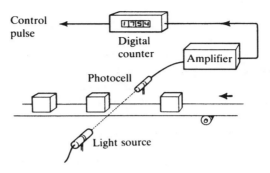

Figure 25.2 *Using a light sensor to count packages*

logic counter circuit. This counts the number of sounds picked up by the microphone and displays this as the total number of items counted.

A more sophisticated method which may be used for larger or fragile objects uses a photocell (such as a photodiode) as shown in Figure 25.2. The photocell is placed opposite a light source so that the objects interrupt the light beam as they pass. The drop in the amount of current passing through the photocell is detected as a fall in the p.d. across a resistor in series with the photocell. The falling p.d. triggers a digital counter. Not only can this count the total number of objects but it can also provide an output signal when a certain number of objects has been counted. The signal may be used to initiate some further action, such as stopping the conveyor belt. A similar method is used to count the number of cars entering a multi-storey car park. With two counters, one placed at the entrance and another at the exit, it is possible to keep a tally of the number of cars in the park and to display warning notices and close the entrance gate whenever the park is full.

Counting manufactured items is related to the task of stock-taking. One way in which electronics helps with this is by use of **bar codes**. A bar code is a form of information that is readable by scanning a printed array of bars. The product bearing the bar code is passed across a window and scanned by a high-intensity, narrowly-focused beam of red light, produced by a laser. In some forms of bar code reader, a stationary hand-held reader is held against the bar coded area and scans the bar code. The reflected light is sensed and the information is translated into a series of time intervals. The bars and the spaces between them are either one, two, three or four units wide. Scan times depend on how fast the bar code is wiped across the window. Times are longer if the scan is not exactly perpendicular to the bars. But it is the *relative* lengths of the times that matter. These are determined by analysing the output from the sensor by a computer program. Numbers are coded according to a system in which each number is seven units wide. For example, the figure 8 is coded by the seven binary digits 0110111. Printed as a bar code, this is represented by a space one unit wide, a bar two units wide, a space one unit wide and a bar three units wide. The codes are such that each can be identified whatever the scanning direction; for instance, there is no figure coded by the reverse of the above code, 1110110. But the inverse code has a meaning; for example, the code 0110111 and its inverse code, 1001000, both represent the figure 8. The application of this is that most bar codes consist essentially of two numbers, each of five or six digits, one specifying the manufacturer, and the other specifying the nature of the goods. One number is coded directly and the other inversely, so it is easy for the computer to know which number is which, whatever direction the code is scanned in.

The bar code also includes a single digit indicating the nature of the goods, for example 9 for all books and magazines. A final digit is used as a check number. This is calculated from the previous numbers, according to a formula. On reading the other numbers the computer calculates what the check number should be. If this does not agree with the check number read from the bar code, it indicates that there has been an error in reading the code, and the operator is warned accordingly.

Bar-code readers enable a company to keep a complete list of stock on its computers, adding items to the stock list as they are manufactured or delivered from another company, and removing items from the stock list as they are dispatched to customers. Computers can do more than simply keep count; they can analyse the flow of goods and produce invoices, forecasts of stocks and other useful information.

Bar codes are used wherever an assortment of goods has to be listed. This includes sales receipts in stores. Almost all supermarkets and many other retail stores have a bar code reader linked to the cash register. Usually, information such as price is not coded. This can be held on a central computer and, once the product has been identified from its bar code, the computer looks up the price and sends it to the checkout till. This allows prices to be changed in the central computer by the retailer should there be price increases or special sales offers.

Measuring flow

Measuring the flow of liquids in pipes with no restriction to the flow of the liquid is achieved by the arrangement shown in Figure 25.3. Two ultrasonic transducers are mounted in the pipe as shown. An automatic switching arrangement makes one transducer act as an emitter of ultrasound while the other acts as a receiver. An instant later their functions are reversed. In this way the equipment measures the time taken for pulses of ultrasound to travel downstream and upstream. The difference between these times is directly proportional to the rate of flow. The time differences are calculated by suitable logical circuitry and the flow rate displayed or recorded. Another technique relies on the Doppler effect, and is suitable for liquids that contain either solid

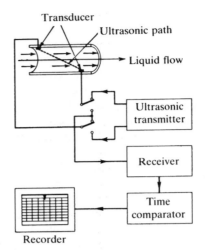

Figure 25.3   *An ultrasonic flowmeter*

particles or bubbles. In this technique the two transducers are placed side by side in the pipe directed against the flow direction. One is a transmitter and the other a receiver as in the usual arrangement for Doppler effect measurement (p. 263). Ultrasound is directed toward the oncoming bubbles of particles, is reflected from them and is detected by the receiving transducer. The beat rate gives a measure of the rate of flow of the liquid.

Electronics is also able to measure the rate of flow of gases. An ingenious gas flow sensor together with the necessary circuit can be fabricated on a silicon chip (Figure 25.4). As the gas flows through the orifice it presses against the springy strip. The faster the flow the greater the pressure and the greater the

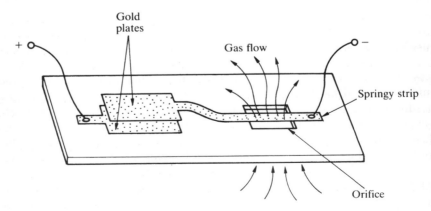

Figure 25.4   *A sensor for measuring the rate of flow of gas*

distance between the gold plates. The attached circuit measures the capacitance between the gold plates and, because this depends on their distance apart, it is proportional to the gas flow. Such a device is very small and could replace the domestic gas meter. Moreover, it lends itself to remote reading, so that instead of the meter reader having to call at the house (to find the occupants away for the day) the meter is read from a central station.

Whether we are measuring the flow of liquids or measuring the flow of traffic, electronics has an application. On today's busy road systems it is essential to companies as well as to the general public that the traffic should be kept moving. Electronics makes many contributions to this. One example is **Trafficmaster**®, which is a system for warning motorists of traffic jams and similar causes of delays on motorways. Infra-red sensors mounted on bridges over the motorways measure the speed of traffic passing beneath them, and the number of vehicles passing. The speeds are averaged for successive three-minute periods and, when the average falls below 30 mph, a signal is sent by radio to the control centre of the organization. Incoming signals are analysed by computer and the results are sent by radio to a pager mounted on the dashboard of the car. The information is presented to the driver in the form of a motorway map showing the points where jams are occurring, and the average vehicle speeds at these points. Sensors are, on average two miles apart, so three adjacent points indicating low-speed traffic implies a tail-back at least four miles long. The driver can also call up text messages explaining the causes of the trouble in more detail. The information is updated every three minutes, giving a continuous assessment of the traffic situation. A system such as this, which works just as well at night and in thick fog, provides information aimed at increasing road safety, cutting costs and travel time, and reducing pollution.

Measuring levels

In many automated industrial processes it is essential to know the level of a liquid or powdered substance stored in a tank. One of several possible techniques is illustrated in Figure 25.5. A source of ultrasonic pulses is directed downward on to the substance in the tank. Part of the ultrasound is reflected from the top surface of the liquid and detected by the sensor. Part is detected after reflection from the bottom of the tank. The difference between the time of arrival of the two reflected pulses is a measure of the depth of the liquid. This technique is equivalent to using a dip-stick but there is one important difference. There is no easy way in which the information obtained by using a dipstick can be passed automatically to an electronic control circuit. By contrast, the time differences measured by the circuit of the ultrasonic equipment may be fed directly to other electronic circuits. The same applies to the other measuring examples quoted in this chapter; all of them are able to pass information directly to a computer or other electronic controller.

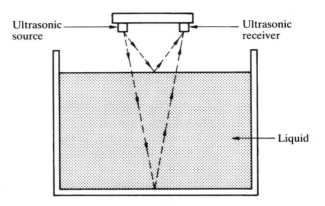

Figure 25.5 *An ultrasonic liquid level detector*

Measuring position

In an automatically-controlled machine, it may be important for the control circuit to 'know' the exact position of certain parts at any given instance. It may be essential to know the position of the work-piece or of a tool such as a drill bit, or that a safety gate has been shut. It is not enough to have a motor or piston that moves a part to a given position. Friction, jamming, a 'spanner in the works' or some other such mishap may have prevented the part from reaching its intended location. A positive measurement of its position is required. One such system is illustrated in Figure 25.6, applicable to a part that slides to and fro in a straight line. A sliding panel with clear and opaque sections is attached to the sliding part. A row of four photodiodes sends infra-red radiation coming from behind the panel. In the position shown in the figure, in which all four sections are clear, infra-red reaches all the photo-

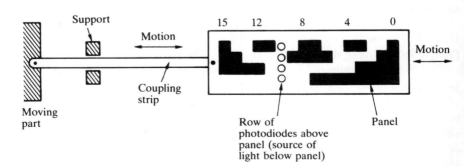

Figure 25.6 *Using the Gray code to determine position*

diodes. The photodiodes are each connected to a logical circuit and in this example all four outputs are high (1111). The panel (and therefore the sliding part) has sixteen positions, numbered from 0 to 15. Since there are four photodiodes, the pattern on the panel could be set out so as to represent the binary numbers from 0000 to 1111 in numerical order. Instead, as can be seen in the figure, the pattern gives 0000 for position 0 (as might be expected) but 1000 for position 1, 1100 for position 2 and so on. Each position is represented by a four-digit code, known as the **Gray code**. The point about the Gray code is that, as the panel moves from one position to the next, only one digit is affected. This helps eliminate a source of error. It is not easy to be sure that the row of photodiodes is so straight and is aligned so precisely that changes in the panel position affect all four of them at exactly the same instant. It is more likely that they are affected in some random order. For example, if positions 7 and 8 were coded as ordinary binary numbers, 0111 and 1000, the change from one to the other might go through this sequence:

	Position				Position
Digit	7	Changing to			8
A	1	1	1	* 0	0
B	1	* 0	0	0	0
C	1	1	1	1	* 0
D	0	0	* 1	1	1

The asterisks indicate the points at which the photodiodes are affected by the motion of the panel. In this example they change in the order BDAC, and generate numbers 5, 13 and 12 while doing so. This gives completely false readings of the position as the panel moves. In moving from positions 7 to 8 with the Gray code, the change is from 1101 to 1100; only digit D is affected. The other digits remain as before, so the order in which they are affected does not matter and there are no spurious intermediate states.

Special ICs are made for converting the Gray code into its binary equivalent. Being digital, the output of this is ideal for sending to any other logic circuit or to a computer for use in automatic control of machinery.

Detecting flaws

In the iron founding industry, ultrasonics has assisted in making quality testing a much more exact task. In earlier days, large iron castings weighing many hundreds of kilograms were sold to customers for machining. This is a process which might take several weeks and toward the end of that time a large hole might be revealed in the block, due to faulty casting. This meant not only a waste of the casting but a serious loss of time. Even though all possible precautions are taken to avoid large flaws, it is impossible to guarantee that they

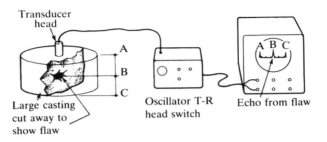

Figure 25.7   *Using ultrasound to check for flaws in metal castings*

will not occur. But castings can be inspected for flaws by using an ultrasonic technique.

The transducer head is coupled to the casting, as shown in Figure 25.7, by using a water-bound cellulose paste. The oscillator is pulsed (as in radar) so that it produces short sharp output pulses at the rate of several hundreds per second, each pulse lasting a few milliseconds. The pulses are fed to the transducer and transmitter through the casting. Echoes from internal surfaces such as the surfaces of flaws return along the path of the beam. These echoes are picked up by the transducer, reconverted to electrical signals and displayed on an oscilloscope. In this way the presence of a flaw shows up as 'blips' on the oscilloscope screen.

The same transducer head acts both as transmitter and receiver. After each transmitted pulse the head is electronically switched over to receive the echo. Figure 25.7 shows a typical set of echoes from a casting. Echo A is caused from the interface between the transducer and the casting. Echo B is caused by the flaw. Echo C is from the lower surface of the casting. The distance between traces gives an indication of the vertical position of the flaw within the casting.

Weighing massive objects

Electronic weighing machines are widely used now in the kitchen, the supermarket checkout and the post office. These mainly depend on a piezo-electric sensor which generates a p.d. proportional to the force exerted on it by the object on the scale-pan. In heavy industry it may be necessary to measure extremely large forces or to weigh vehicles and other massive objects. One way of doing this is by using a **load cell**. A typical load cell, used for measuring compressive forces, such as weight consists of a strain gauge (p. 109) attached to a stout metal cylinder (Figure 25.8). The cylinder supports a platform on which the vehicle or other heavy object is placed. The weight of the object

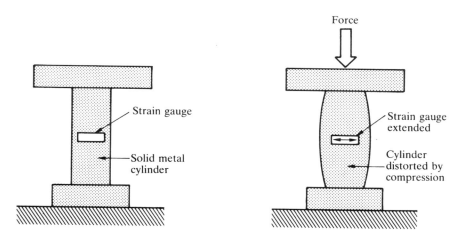

Figure 25.8   *Using a load cell to measure a large compressive force*

distorts the cylinder, and this strain is measured by a circuit connected to the strain gauge. The result is read out in terms of force or weight.

Measuring high temperatures

The temperatures in furnaces usually are so high that any thermal sensor placed directly in the furnace is immediately destroyed. Instead of using a thermometer we use a **radiation pyrometer** to sense the type and amount of radiation being emitted from a furnace and use this to estimate temperature. Measuring the temperature of very hot objects calls for similar instruments. Pyrometers are based on various sensors, including thermopiles, thermistors and pyroelectric devices, all described elsewhere in this book. A pyrometer often has a telescope built into it, so that it may be directed accurately at the open furnace door or the hot object. This may incorporate a dichroic mirror to allow visible light to pass through to the telescope but to reflect infra-red onto the sensor. Some pyrometers work on the chopped-beam principle. A bladed 'fan' in the optical system alternately exposes the sensor to radiation from the furnace and radiation from the filament of a standard lamp. The current through the filament is adjusted automatically until the sensor receives equal amounts of radiation from either source. In other words the output from the sensor is steady, not alternating. The temperature of the furnace is then found by referring to calibration charts.

A **bolometer** does not sense infra-red directly, but measures its heating effect on a thermistor. The instrument has a matched pair of thermistors, one of which is exposed to the radiation, the other of which is shielded. The exposed thermistor is blackened to assist the absorption of infra-red. The purpose of the shielded thermistor is to compensate for changes in ambient temperature. The resistances of the thermistors are compared using a bridge circuit (p. 60). The output of the bridge circuit is proportional to the amount of radiation falling on the exposed thermistor, and hence to the temperature of the hot object.

Digital printing

Previously, any technique for printing involved a heavy initial expenditure of both time and money for the preparation of printing plates. In colour printing this was preceded by the making of colour separation negatives or positives. Nowadays, these stages may be by-passed by using a computer.

Digital printing is an ideal technique for printing a small number of copies (up to about 20) of large-size colour pictures. It is specially suitable for producing posters and displays for exhibitions. All the initial work is done on the computer and usually begins with a small colour print being scanned and stored on disk. Colour transparencies and drawings may also be used and can be combined with text. It is also possible for original images to be received directly over a computer network. The next stage is to edit the image using conventional graphics software. This may include cropping the image, adding textures, altering its colour balance, deliberately distorting its shape, removing blemishes and adding superimposed text in a wide range of styles. At this point the edited image is ready for printing.

The image is printed out using a colour plotter under computer control. It prints on paper of any feasible width, up to a metre or more wide. The image is printed in five runs with highly accurate superposition: registration marks, cyan, magenta, yellow and black. Together these build up a full-colour image as in conventional printing, but the computer software decides the density for each colour at each point in the image. The software allows an exceptionally high degree of magnification without graininess or loss of definition. Special program routines operate to produce photographic quality whatever the scale of enlargement.

The printing is done by a process similar to that used in a photocopier. At each run the paper is charged electrostatically according to the required image. It then passes through a toner bath, where the pigment is attracted to the paper in the charged areas. The solvent dries almost instantly, leaving the pigment adhering firmly to the paper.

Moisture content

Very often, as in the example of the gas flow sensor mentioned earlier, it is easy to use a basic electronic quantity for measuring what may be a quite complex

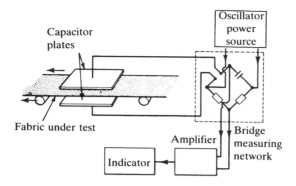

Figure 25.9 *Monitoring the moisture content of fabric by measuring capacitance*

attribute of a product. An example of this is using capacitance to measure the moisture content of fabric (Figure 25.9). The fabric is passed over rollers between two metal plates which act as the plates of a capacitor. The fabric is thus the equivalent of the dielectric of the capacitor. The capacitance is proportional to the moisture content of the fabric. This may be measured using a bridge circuit (p. 60). If the capacitance is found to be outside acceptable limits the circuit may operate an alarm or may automatically take an action such as raising the temperature of driers.

In agriculture, a similar method is used in grain driers. A large specially shaped capacitor is situated in the ducts in which grain is passed through the drier (Figure 25.10). Changes in capacitance produce signals which are used to control the rate at which the grain passes through the drier, to ensure that it is evenly dried.

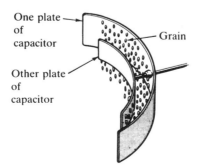

Figure 25.10 *A capacitor being used in a grain drier to measure moisture content*

Automation

Electronic techniques have made possible great advances in the automation of industrial processing. In the chemical industry, for example, some large refining plant are run for months with only a handful of staff, most of the processes being automated.

Figure 25.11 shows in outline a basic automatic control system. Suppose we wish a conveyor belt to travel at a certain speed. This is referred to in Figure 25.11 as the 'operation under control'. The belt is driven by an electric motor. To this is linked a device such as a Hall-effect tachometer (p. 111). The output of this is proportional to the speed of the motor. This signal (which could be either an analogue signal or a digital one) is compared with a reference signal, using a comparator circuit. The reference signal, is either an analogue p.d. or a digital alternating voltage. It is set manually or by computer to make the motor run at the correct speed. The output of the comparator is an error signal, depending on the difference between the actual speed of the motor and its correct speed. The error signal, after amplification, is used to control the power delivered to the motor so that its speed is automatically regulated. This is called **closed-loop feedback control**. Information on the actual operation under control is fed back and used to adjust it automatically. There is, of course, some slight error in the system because deviation can not be corrected until after it has occurred, but this error can be kept very small by careful design of the system.

Feedback control can be devised for a wide range of quantities including temperature, humidity flow, pressure and weight. Feedback control of position may be accomplished with the Gray-code sensor described on p. 280. Also, if the reference signal is programmed to change at different stages in an operation, the controlled device can be made to perform a sequence of operations automatically. A program can be provided on disk for a computer and used in

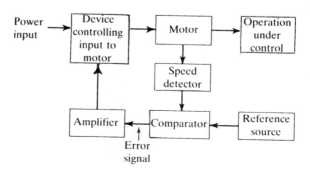

Figure 25.11   *Block diagram of the elements of a closed-loop automatic control system*

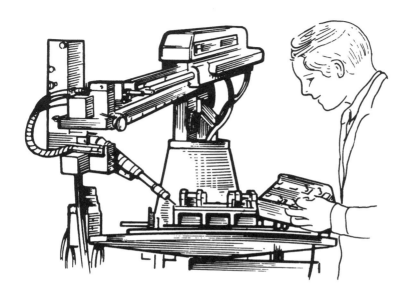

Figure 25.12 *A factory robot. The engineer is 'teaching' the machine to carry out a set routine, after which it will repeat the routine automatically*

the control, for example of machine tools. Standard control 'languages' have been devised for such operations, for example p.c.b.s (p. 134) are drilled by numerically controlled drilling machines. Using the standard format, the computer disk specifies the size and position of all the holes in a board and the machine drills the holes automatically.

Electronic systems of automation relieve human operators of the tedious tasks of measuring, assessing and correcting equipment and, moreover, can operate at much faster speeds. Many factory jobs such as painting, welding and assembling of machinery are performed by robots. Some of these are taken through the required routines by a human operator (Figure 25.12). After one such 'training' session the robot 'remembers' the movements required (for example, to spray-paint the body of a car) and repeats the action on every subsequent occasion without the need for human guidance. The essential feature of robots is that they are flexible; they can be programmed to perform a variety of different tasks, so that the same robot can be used in different parts of the production line, or can be re-programmed when the product design is altered.

At this point in our account of automation we are going beyond the province of electronics. The action of robots is controlled by software, and their seeming skill is the result of the skill of the programmer. Their reliability is the result of the skill of the engineer. These are aspects which it would be out of place to discuss here, but the electronic sensors used by the robots are those described in earlier chapters in this book. The electronic circuits of the robot are based on the transistors and logic gates we have already described. Those who have read this far in the book will find nothing new in the electronic aspects of these apparently human machines.

26
Electronics and the future

Many times in this book we have illustrated the increasing impact of electronics on our lives. It seems safe to predict that this tendency will continue. The proposed 'Information superhighway' is an indicator of this trend. We have virtually instant access to any information on a wide scale and at an affordable cost. Soon a small box sitting on top of the domestic TV receiver will put the owner in contact with an international computer network. This will be able to provide not only the electronic version of a video library, so that we can view any film at any time or receive personalized news broadcasts dealing with topics that interest us most, but will also enable us to shop, carry out financial transactions, make travel bookings and in general interact with people and enterprises throughout the World. Already, large companies are publishing their sales catalogues in the form of CD-ROMs, and it will be possible to down-load these over the network, and order and pay for the goods – all without leaving home. In the field of education we may expect to see more courses of the kind already being televised by the Open University, but with the backup of interactive tutorials in which student and tutor communicate directly by way of the network. The skills of gifted teachers will be available to a much wider audience than they are today. Working from home will become more widespread, which will benefit many groups of persons. For the ambitious it will give increasing job mobility, for it will no longer be necessary to move house in order take up a job with a company situated in another town or even in another country. Mothers will be able to work at home and still give their young children the care they need. The physically handi-capped will have greater opportunities for employment. These are some of the benefits that the expansion of electronic communications may offer. There may be problems too, such as safeguards against invasion of personal privacy, and the maintenance of the security of data. It may be necessary for society to adjust some of its practices and attitudes if the greatest possible benefits are to be realized.

For those who are obliged to or prefer to leave their homes to work, elec-tronics will continue to provide increasing benefits; more comfortable and safer transport to work or school, more congenial work conditions, greater safety at work, and probably shorter working hours.

But what of future developments in electronics itself? Here the most likely forecast can be summed up in three words: smaller, faster, and smarter.

There are two aspects of the forecast that electronics will become smaller. The first applies to electronically-based equipment such as telephones, test-meters and computing devices. By the increased use of VLSI (p. 142) and by the miniaturization of components, the physical size of equipment will be further reduced. There is, of course, a lower limit set, for example, by the minimum convenient size for a keyboard. This limit has already been reached with certain types of equipment. But there is still scope for miniaturization in other directions. Electronic circuits need a power supply and research into new techniques of battery construction are aimed at high energy density – obtaining maximum power from cells of minimum size. These will allow compact equipment to become even more compact, or allow existing equipment to run for longer before the battery is exhausted. Already we have a diminutive computer system, complete with processor, RAM, a high-level programming language in its ROM, input ports, output ports, and space for installing additional circuitry, all on a circuit board little larger than a postage stamp. One might wonder when we shall have wrist-watch-sized colour TV receivers like those featured in science fiction films.

The more significant aspect of size reduction is the miniaturization of the components themselves. Much research is aimed at making transistors smaller than ever, with the aim of packing more and more of them on the chip. New semiconductors, such as those based on gallium arsenide, promise to give us smaller, more densely packed transistors, leading to a new phase of larger-scale integration. Small size is not the sole aim of these researches; a densely-packed circuit has shorter connections between its components, which means that signals pass from one to the other more quickly. Operating speed can be significantly increased. A parallel effect arises because smaller transistors have lower capacitance and switch states more rapidly. Power consumption is reduced too. All of this encourages the use of microprocessors and other complex ICs in electronic systems, leading to an overall increase in 'smartness' of all our electronic equipment. This trend is set to continue, and our smaller-sized circuits also will be both faster and smarter. The three qualities go hand-in-hand.

Apart from enhancing the performance of semiconductors and looking for new ones, researchers are examining other ways of building components and circuits based on new technologies. At present, the semiconductors in use in electronics are elements (silicon, germanium) or inorganic compounds (cadmium sulphide, gallium arsenide). Now the semiconducting properties of some of the organic compounds are under investigation. For example, it has been shown that a film of phenylene vinylene emits a greenish yellow light when a p.d. of 1.5 V is applied across it. This organic LED has very low efficiency, but research is showing ways of increasing this. Advances in this field will no doubt lead to the discovery of new organic semiconductors and eventually to a range of components based on this new technology.

Another branch of electronic research seeks to produce the smallest possible components, consisting of single molecules. The behaviour of certain kinds of molecules is analogous to the behaviour of semiconductor materials. It is feasible that we could design and build molecules that will perform the same functions as transistors and other semiconducting devices. Molecular electronics, as it is called, would allow circuits to be made very much smaller than is possible with techniques for fabricating integrated circuits from semiconductors. One of the biggest problems will be connecting such one-molecule devices to external circuitry. This difficulty is one of the more important impediments to the development of molecular devices. At present, research is concentrating on devices built not from single molecules but from films of long molecules lying parallel to each other. Already it has been shown that such groups of molecules pass current more readily in one direction than in the other. The film acts as a diode. Further, the action seems to be sensitive to light intensity, giving us a photodiode. The next step is to assemble molecular photodiodes to give a structure which acts as a transistor. The aim is to make diodes and transistors consisting of a single molecule, but many hurdles have to be overcome before this is achieved.

A promising approach to molecular electronics is a molecule which performs the logical AND operation (p. 187). This was developed by Dr. A P de Silva and his colleagues at Queens University, Belfast. When the molecule is irradiated by UV it normally makes no response. If it is irradiated in the presence of ions of hydrogen, it likewise makes no response; neither does it respond in the presence of ions of sodium. But if both hydrogen and sodium ions are present, it fluoresces with a blue light when irradiated. The reaction occurs with the molecule and ions in solution showing that the operation is a purely chemical one. It remains to link the reaction to the real world by some kind of interface.

In the decades following the invention of transistors there was a period of exploitation of their properties. In the seventies, new designs of integrated circuits appeared almost daily. Within a few years, the manufacturers had produced ICs to undertake almost all of the special functions that we could ever need. More recently, there have been fewer really novel ICs, most of the new designs being merely improvements on the old ones. Faster action, lower power consumption, and enhanced ability to work in hostile environments have been the main directions of improvement. Very few really new logic circuits appear nowadays. If anyone needs a new logic function, which would be a complex one since all the simple ones are already provided for, the approach is to use an IC with programmable logic gates that can be tailored to fit the requirements.

Programmable logic ICs are one instance of the increasing trend to replace hardware by software. Now that microprocessors and microcontrollers are so versatile and cheap, this course is usually the least expensive and requires the least effort. If there is a function to be performed and there is no IC ready-made to perform it, take a general-purpose microcontroller, equip it with the necessary electronic sensors and output devices, then program it to perform that function. Once this stratum of electronic circuit design has been reached there is less need for new circuit designs and new ICs.

Software is taking over many of the functions formerly fulfilled by complex electronic circuits. For example, instead of using capacitors and inductors to build an audio filter, we can perform the same task by letting a computer program operate in real time on the digitized audio signal, and with superior performance.

Recent developments in computer programming techniques have lent impetus to this trend. One factor is the development of 'fuzzy logic'. Instead of a logic circuit with an absolute truth table giving true/false outputs for all possible combinations of inputs, we have programs which calculate the relative probabilities of possible logical outcomes. This corresponds more closely to the situations in which we find ourselves in real life. The development of neural networks is another new application of computers. A neural network is a technique for operating a computer by a program which models the behaviour of a nerve system. Like a nerve system, it is able to learn by trial and error and to gradually adopt suitable ways of handling and responding to data supplied to it. This application of computer power is still in its infancy but already it has achieved striking successes. Take for example the neural network which has been configured as an intelligent system to recognise dirty, damaged, or tilted motor registration plates. This has obvious practical applications, including the apprehension of motorway driving offenders. An image of the number plate of a passing vehicle is scanned by a TV camera and the digitized data is fed to a computer. The neural network of the computer is triggered automatically by the yellow or grey image of a number plate. The number, indistinct though it may be, is recognized within 0.25 s, with a 90% success rate.

Another practical application of neural networks was in identifying the risk factors leading to fires in crashed aircraft. The computer was provided with records of 17 000 aircraft crashes, some of which resulted in fires and some of which did not. The neural network was asked to examine the data and look for correlations between the circumstances of the crash and the incidence of fire. Looking through the data, the neural network was able to detect weak correlations within the data. When it had completed its analysis it was tested on data related to 2000 crashes. It correctly predicted fires in 70% of cases. Attempting to predict fire using conventional statistical techniques is successful in only 10% of cases.

From the above, it seems that the principal contribution of electronics in the foreseeable future will be to provide the means for building up massive computing power. But whatever the future holds, it is clear that electronics will continue to play a major part in all of our lives.

Index